# Biological Data Integration

SCIENCES

*Computer Science*, Field Directors –
Valérie Berthé and Jean-Charles Pomerol

*Bioinformatics*, Subject Head –
Anne Siegel and Hélène Touzet

# Biological Data Integration

## Computer and Statistical Approaches

*Coordinated by*
Christine Froidevaux
Marie-Laure Martin-Magniette
Guillem Rigaill

WILEY

First published 2023 in Great Britain and the United States by ISTE Ltd and John Wiley & Sons, Inc.

Apart from any fair dealing for the purposes of research or private study, or criticism or review, as permitted under the Copyright, Designs and Patents Act 1988, this publication may only be reproduced, stored or transmitted, in any form or by any means, with the prior permission in writing of the publishers, or in the case of reprographic reproduction in accordance with the terms and licenses issued by the CLA. Enquiries concerning reproduction outside these terms should be sent to the publishers at the under mentioned address:

ISTE Ltd
27-37 St George's Road
London SW19 4EU
UK

www.iste.co.uk

John Wiley & Sons, Inc.
111 River Street
Hoboken, NJ 07030
USA

www.wiley.com

© ISTE Ltd 2023

The rights of Christine Froidevaux, Marie-Laure Martin-Magniette and Guillem Rigaill to be identified as the authors of this work have been asserted by them in accordance with the Copyright, Designs and Patents Act 1988.

Any opinions, findings, and conclusions or recommendations expressed in this material are those of the author(s), contributor(s) or editor(s) and do not necessarily reflect the views of ISTE Group.

Library of Congress Control Number: 2023932774

British Library Cataloguing-in-Publication Data
A CIP record for this book is available from the British Library
ISBN 978-1-78945-030-9

ERC code:
LS2 Genetics, 'Omics', Bioinformatics and Systems Biology
 LS2_12 Bioinformatics
 LS2_13 Computational biology
 LS2_14 Biostatistics
 LS2_15 Systems biology

# Contents

**Preface** . . . . . . . . . . . . . . . . . . . . . . . . . . . . . . . . . . . xi
Christine FROIDEVAUX, Marie-Laure MARTIN-MAGNIETTE and
Guillem RIGAILL

**Part 1. Knowledge Integration** . . . . . . . . . . . . . . . . . . . . . . 1

**Chapter 1. Clinical Data Warehouses** . . . . . . . . . . . . . . . . . . . 3
Maxime WACK and Bastien RANCE

    1.1. Introduction to clinical information systems and biomedical
    warehousing: data warehouses for what purposes? . . . . . . . . . . . . . 3
        1.1.1. Warehouse history . . . . . . . . . . . . . . . . . . . . . . . . 4
        1.1.2. Using data warehouses today . . . . . . . . . . . . . . . . . . 4
    1.2. Challenge: widely scattered data . . . . . . . . . . . . . . . . . . . 5
    1.3. Data warehouses and clinical data . . . . . . . . . . . . . . . . . . 6
        1.3.1. Warehouse structures . . . . . . . . . . . . . . . . . . . . . . 6
        1.3.2. Warehouse construction and supply . . . . . . . . . . . . . . . 11
        1.3.3. Uses . . . . . . . . . . . . . . . . . . . . . . . . . . . . . . . 11
    1.4. Warehouses and omics data: challenges . . . . . . . . . . . . . . . . 15
        1.4.1. Challenges of data volumetry and structuring omic data . . . . . 16
        1.4.2. Attempted solutions . . . . . . . . . . . . . . . . . . . . . . . 17
    1.5. Challenges and prospects . . . . . . . . . . . . . . . . . . . . . . . 18
        1.5.1. Toward general-purpose warehouses . . . . . . . . . . . . . . . 18
        1.5.2. Ethical dimension of the implementation and the use of
        warehouses . . . . . . . . . . . . . . . . . . . . . . . . . . . . . . . 19
        1.5.3. Origin and reproducibility . . . . . . . . . . . . . . . . . . . . 19
        1.5.4. Data quality . . . . . . . . . . . . . . . . . . . . . . . . . . . 20
        1.5.5. Data warehousing federation and data sharing . . . . . . . . . . 21
    1.6. References . . . . . . . . . . . . . . . . . . . . . . . . . . . . . . 21

## Chapter 2. Semantic Web Methods for Data Integration in Life Sciences  25
Olivier DAMERON

    2.1. Data-related requirements in life sciences . . . . . . . . . . . . . . . . . 26
        2.1.1. Databases for the life sciences . . . . . . . . . . . . . . . . . . . . 26
        2.1.2. Requirements . . . . . . . . . . . . . . . . . . . . . . . . . . . . 27
        2.1.3. Common approaches: InterMine and BioMart . . . . . . . . . . . 30
    2.2. Semantic Web . . . . . . . . . . . . . . . . . . . . . . . . . . . . . . . 31
        2.2.1. Techniques . . . . . . . . . . . . . . . . . . . . . . . . . . . . . 32
        2.2.2. Implementation . . . . . . . . . . . . . . . . . . . . . . . . . . . 42
    2.3. Perspectives . . . . . . . . . . . . . . . . . . . . . . . . . . . . . . . . 43
        2.3.1. Facilitating appropriation to users . . . . . . . . . . . . . . . . . 43
        2.3.2. Facilitating the appropriation by software programs: FAIR data . . . . . . . . . . . . . . . . . . . . . . . . . . . . . . . . . 44
        2.3.3. Federated queries . . . . . . . . . . . . . . . . . . . . . . . . . . 45
    2.4. Conclusion . . . . . . . . . . . . . . . . . . . . . . . . . . . . . . . . 46
    2.5. References . . . . . . . . . . . . . . . . . . . . . . . . . . . . . . . . 47

## Chapter 3. Workflows for Bioinformatics Data Integration  53
Sarah COHEN-BOULAKIA and Frédéric LEMOINE

    3.1. Introduction . . . . . . . . . . . . . . . . . . . . . . . . . . . . . . . . 53
    3.2. Bioinformatics data processing chains: difficulties . . . . . . . . . . . 54
        3.2.1. Designing a data processing chain . . . . . . . . . . . . . . . . . 55
        3.2.2. Analysis execution and reproducibility . . . . . . . . . . . . . . 56
        3.2.3. Maintenance, sharing and reuse . . . . . . . . . . . . . . . . . . 58
    3.3. Solutions provided by scientific workflow systems . . . . . . . . . . . 59
        3.3.1. Fundamentals of workflow systems . . . . . . . . . . . . . . . . 59
        3.3.2. Workflow systems . . . . . . . . . . . . . . . . . . . . . . . . . 64
    3.4. Use case: RNA-seq data analysis . . . . . . . . . . . . . . . . . . . . . 69
        3.4.1. Study description . . . . . . . . . . . . . . . . . . . . . . . . . . 69
        3.4.2. From data processing chain to workflows . . . . . . . . . . . . . 72
        3.4.3. Data processing chains implemented as workflows: conclusion . . 75
    3.5. Challenges, open problems and research opportunities . . . . . . . . . 77
        3.5.1. Formalizing workflow development . . . . . . . . . . . . . . . . 77
        3.5.2. Workflow testing . . . . . . . . . . . . . . . . . . . . . . . . . . 78
        3.5.3. Discovering and sharing workflows . . . . . . . . . . . . . . . . 79
    3.6. Conclusion . . . . . . . . . . . . . . . . . . . . . . . . . . . . . . . . 80
    3.7. References . . . . . . . . . . . . . . . . . . . . . . . . . . . . . . . . 81

**Part 2. Integration and Statistics** . . . . . . . . . . . . . . . . . . . . . . . . 87

**Chapter 4. Variable Selection in the General Linear Model:
Application to Multiomic Approaches for the Study of Seed Quality**  89
Céline LÉVY-LEDUC, Marie PERROT-DOCKÈS, Gwendal CUEFF and
Loïc RAJJOU

    4.1. Introduction . . . . . . . . . . . . . . . . . . . . . . . . . . . . . . . . 90
    4.2. Methodology . . . . . . . . . . . . . . . . . . . . . . . . . . . . . . . 93
        4.2.1. Estimation of the covariance matrix $\Sigma_q$ . . . . . . . . . . . . . . 93
        4.2.2. Estimation of $\mathcal{B}$ . . . . . . . . . . . . . . . . . . . . . . . . . . . . 96
    4.3. Numerical experiments . . . . . . . . . . . . . . . . . . . . . . . . . 99
        4.3.1. Statistical performance . . . . . . . . . . . . . . . . . . . . . . . 99
        4.3.2. Numerical performance . . . . . . . . . . . . . . . . . . . . . . . 100
    4.4. Application to the study of seed quality . . . . . . . . . . . . . . . . . 103
        4.4.1. Metabolomics data . . . . . . . . . . . . . . . . . . . . . . . . . 104
        4.4.2. Proteomics data . . . . . . . . . . . . . . . . . . . . . . . . . . . 105
    4.5. Conclusion . . . . . . . . . . . . . . . . . . . . . . . . . . . . . . . . 108
    4.6. Appendices . . . . . . . . . . . . . . . . . . . . . . . . . . . . . . . . 108
        4.6.1. Example of using the package `MultiVarSel` for metabolomic
        data analysis . . . . . . . . . . . . . . . . . . . . . . . . . . . . . . . . 108
        4.6.2. Example of using the package `MultiVarSel` for proteomic
        data analysis . . . . . . . . . . . . . . . . . . . . . . . . . . . . . . . . 110
    4.7. Acknowledgments . . . . . . . . . . . . . . . . . . . . . . . . . . . . 113
    4.8. References . . . . . . . . . . . . . . . . . . . . . . . . . . . . . . . . 113

**Chapter 5. Structured Compression of Genetic Information
and Genome-Wide Association Study by Additive Models** . . . . . . 117
Florent GUINOT, Marie SZAFRANSKI and Christophe AMBROISE

    5.1. Genome-wide association studies . . . . . . . . . . . . . . . . . . . . 118
        5.1.1. Introduction to genetic mapping and linkage analysis . . . . . . . 118
        5.1.2. Principles of genome-wide association studies . . . . . . . . . . 119
        5.1.3. Single nucleotide polymorphism . . . . . . . . . . . . . . . . . . 120
        5.1.4. Disease penetrance and *odds ratio* . . . . . . . . . . . . . . . . 122
        5.1.5. Single marker analysis . . . . . . . . . . . . . . . . . . . . . . . 124
        5.1.6. Multi-marker analysis . . . . . . . . . . . . . . . . . . . . . . . 126
    5.2. Structured compression and association study . . . . . . . . . . . . . 132
        5.2.1. Context . . . . . . . . . . . . . . . . . . . . . . . . . . . . . . . 132
        5.2.2. New structured compression approach . . . . . . . . . . . . . . 133
    5.3. Application to ankylosing spondylitis (AS) . . . . . . . . . . . . . . . 142
        5.3.1. Data . . . . . . . . . . . . . . . . . . . . . . . . . . . . . . . . . 142
        5.3.2. Predictive power evaluation . . . . . . . . . . . . . . . . . . . . 143

5.3.3. Manhattan diagram . . . . . . . . . . . . . . . . . . . . . . . . . 144
5.3.4. Estimation for the most significant SNP aggregates . . . . . . . . 144
5.4. Conclusion . . . . . . . . . . . . . . . . . . . . . . . . . . . . . . . 146
5.5. References . . . . . . . . . . . . . . . . . . . . . . . . . . . . . . . 146

**Chapter 6. Kernels for Omics** . . . . . . . . . . . . . . . . . . . . . . 151
Jérôme MARIETTE and Nathalie VIALANEIX

6.1. Introduction . . . . . . . . . . . . . . . . . . . . . . . . . . . . . . . 152
6.2. Relational data . . . . . . . . . . . . . . . . . . . . . . . . . . . . . 153
    6.2.1. Data described by the kernel . . . . . . . . . . . . . . . . . . 153
    6.2.2. Data described by a general (dis)similarity measure . . . . . . . 155
6.3. Exploratory analysis for relational data . . . . . . . . . . . . . . . . 158
    6.3.1. Kernel clustering . . . . . . . . . . . . . . . . . . . . . . . 158
    6.3.2. Kernel principal component analysis . . . . . . . . . . . . . . 161
    6.3.3. Kernel self-organizing maps . . . . . . . . . . . . . . . . . . 163
    6.3.4. Limitations of relational methods . . . . . . . . . . . . . . . 166
6.4. Combining relational data . . . . . . . . . . . . . . . . . . . . . . . 168
    6.4.1. Data integration in systems biology . . . . . . . . . . . . . . 168
    6.4.2. Kernel approaches in data integration . . . . . . . . . . . . . 169
    6.4.3. A consensual kernel . . . . . . . . . . . . . . . . . . . . . . 172
    6.4.4. A parsimonious kernel that preserves the topology of the
    initial data . . . . . . . . . . . . . . . . . . . . . . . . . . . . . 173
    6.4.5. A complete kernel preserving the topology of the initial data . . . 175
6.5. Application . . . . . . . . . . . . . . . . . . . . . . . . . . . . . . . 176
    6.5.1. Loading Tara Ocean data . . . . . . . . . . . . . . . . . . . . 176
    6.5.2. Data integration by kernel approaches . . . . . . . . . . . . . 177
    6.5.3. Exploratory analysis: kernel PCA . . . . . . . . . . . . . . . . 179
6.6. Session information for the results of the example . . . . . . . . . . 186
6.7. References . . . . . . . . . . . . . . . . . . . . . . . . . . . . . . . 188

**Chapter 7. Multivariate Models for Data Integration and
Biomarker Selection in 'Omics Data** . . . . . . . . . . . . . . . . . . . 195
Sébastien DÉJEAN and Kim-Anh LÊ CAO

7.1. Introduction . . . . . . . . . . . . . . . . . . . . . . . . . . . . . . . 195
7.2. Background . . . . . . . . . . . . . . . . . . . . . . . . . . . . . . . 197
    7.2.1. Mathematical notations . . . . . . . . . . . . . . . . . . . . . 197
    7.2.2. Terminology . . . . . . . . . . . . . . . . . . . . . . . . . . 198
    7.2.3. Multivariate projection-based approaches . . . . . . . . . . . . 198
    7.2.4. A criterion to maximize specific to each methodology . . . . . . 199
    7.2.5. A linear combination of variables to reduce the dimension
    of the data . . . . . . . . . . . . . . . . . . . . . . . . . . . . . . 199

|     7.2.6. Identifying a subset of relevant molecular features . . . . . . . . . 200
|     7.2.7. Summary . . . . . . . . . . . . . . . . . . . . . . . . . . . . . . 200
| 7.3. From the biological question to the statistical analysis . . . . . . . . . . 201
|     7.3.1. Exploration of one dataset: PCA . . . . . . . . . . . . . . . . . . 201
|     7.3.2. Classify samples: projection to latent structure discriminant
|     analysis . . . . . . . . . . . . . . . . . . . . . . . . . . . . . . . . . 206
|     7.3.3. Integration of two datasets: projection to latent structure and
|     related methods . . . . . . . . . . . . . . . . . . . . . . . . . . . . . 210
|     7.3.4. Integration of several datasets: multi-block approaches . . . . . . 215
| 7.4. Graphical outputs . . . . . . . . . . . . . . . . . . . . . . . . . . . . . 220
|     7.4.1. Individual plots . . . . . . . . . . . . . . . . . . . . . . . . . . 220
|     7.4.2. Variable plots . . . . . . . . . . . . . . . . . . . . . . . . . . . 221
| 7.5. Overall summary . . . . . . . . . . . . . . . . . . . . . . . . . . . . . 222
| 7.6. Liver toxicity study . . . . . . . . . . . . . . . . . . . . . . . . . . . . 223
|     7.6.1. The datasets . . . . . . . . . . . . . . . . . . . . . . . . . . . . 223
|     7.6.2. Biological questions and statistical methods . . . . . . . . . . . . 223
|     7.6.3. Single dataset analysis . . . . . . . . . . . . . . . . . . . . . . . 224
|     7.6.4. Integrative analysis . . . . . . . . . . . . . . . . . . . . . . . . 231
| 7.7. Conclusion . . . . . . . . . . . . . . . . . . . . . . . . . . . . . . . . 238
| 7.8. Acknowledgments . . . . . . . . . . . . . . . . . . . . . . . . . . . . . 238
| 7.9. Appendix: reproducible R code . . . . . . . . . . . . . . . . . . . . . . 239
|     7.9.1. Toy examples . . . . . . . . . . . . . . . . . . . . . . . . . . . 239
|     7.9.2. Liver toxicity . . . . . . . . . . . . . . . . . . . . . . . . . . . 243
| 7.10. References . . . . . . . . . . . . . . . . . . . . . . . . . . . . . . . . 247

**List of Authors** . . . . . . . . . . . . . . . . . . . . . . . . . . . . . . . . . 251

**Index** . . . . . . . . . . . . . . . . . . . . . . . . . . . . . . . . . . . . . . 255

# Preface

**Christine FROIDEVAUX[1], Marie-Laure MARTIN-MAGNIETTE[2,3] and Guillem RIGAILL[2,4]**

[1]*Université Paris-Saclay, CNRS, LISN, Orsay, France*
[2]*IPS2, Université Paris-Saclay, CNRS, INRAE, Université d'Évry, Université Paris Cité, Gif-sur-Yvette, France*
[3]*MIA Paris-Saclay, Université Paris-Saclay, AgroParis Tech, INRAE, France*
[4]*LaMME, Université Paris-Saclay, CNRS, Université d'Évry, Évry-Courcouronnes, France*

## P.1. Introduction

The study of biological data has undergone fundamental changes in recent years. Firstly, the volume of these data has dramatically increased due to new high-throughput techniques for experiments. Secondly, remarkable advances reached in both computational and statistical analysis methods and in infrastructures have made processing these large datasets possible. These data should then be integrated, that is, their complementarity exploited with the prospect of advancing biological knowledge. Using data integration to allow the most exhaustive analysis possible thus represents a major challenge in biology.

This book intends to address research studies in biological data science with a pedagogical approach, focusing first on computational approaches to biological data integration and then on statistical approaches to omics data integration.

## P.2. Computer-based approaches to biological data integration

### P.2.1. *Challenges of biological knowledge integration*

Biological knowledge has given rise to new fields of application: beyond integrative and systems biology, it is valuable for health and the environment. In particular, the linking of omics data with knowledge of pathologies and clinical data has led to the emergence of precision medicine, which holds tremendous promise for individual health. However, to achieve it, we need to be able to analyze all the knowledge available in an integrated way.

Life sciences data integration must face several difficulties: in addition to the fact that they are massive (Big Data), they are heterogeneous (very varied formats), dispersed (they are found in many databases), presenting various granularities (genomic data or pathology information) and of very variable quality (databases do not all grant the same guarantee of verification (*curation*)).

Unlike other application areas where the integration process is based on the identification of concepts structured in ontologies and on which data matching is performed, biological data integration proceeds by reconciling data using algorithmic, learning and statistical approaches. This integration increasingly attempts to put the human being at the center of the process.

### P.2.2. *Computer-based solutions*

A new paradigm has emerged: the procedure no longer consists of two distinct phases, where the first phase aimed at gathering data distributed through different databases and integrating them, while the second performed analysis on the integrated data. The two phases are intertwined: integration is used for analysis, which in turn is the basis for better integration.

A number of data warehouses have been developed to gather in an integrated, that is to say, structured, coherent and complementary manner fragmented data related to the same biology field. The constitution of these warehouses is accompanied by data querying methods such that their analysis is made possible. These data can be annotated with conceptual terms derived from ontologies, which make it possible to keep track of the deep knowledge associated with them. Ontologies not only allow enriching knowledge with annotations but also to reason about this knowledge. They are at the heart of the Semantic Web, which aims at a fine-grained representation of data to facilitate the automatic integration and interpretation of the data (Chen et al. 2012).

Finally, the analyses performed on the data use a multitude of very different tools. The data processing procedure that makes use of a sequence of several tools one after another, called *workflow*, is becoming a fundamental part of data analysis and is at the heart of the paradigm shift mentioned in the introduction. Designing and executing these bioinformatics data processing chains are important issues.

### P.2.3. *Presentation of the first part*

Chapter 1 introduces data warehouses for the life sciences, focusing on clinical data. Chapter 2 introduces Semantic Web concepts and techniques for omics data integration. Finally, Chapter 3 exposes bioinformatics problems and solutions for designing and executing scientific workflows.

These chapters underline the close relationships between good integration and the FAIR (*Findable, Accessible, Interoperable, Reproducible*) data principles and insist on the importance of data provenance (Zheng et al. 2015). They point out the ethical challenges implied by the protection of stored personal data, especially in the health field, in connection with the security of computer systems.

Throughout these chapters, the reader will be able to see how, in terms of data integration, advances in computational research benefit the life sciences, and how wider adoption of computational methods could benefit them even more so. Conversely, the life sciences offer a tremendous field of investigation for the development of innovative computational methods.

## P.3. Statistical approaches to omics data integration

### P.3.1. *Integration statistical challenges*

Omics data integration is a very broad topic: it is very difficult to accurately define its contours. Our vision of omics data integration is quite close to the one presented by Ritchie et al. (2015):

> [...] (multi)-omics information integration in a meaningful manner to provide a more complete analysis of a biological point of interest.

This definition emphasizes the objectives of integration. The analysis must make sense, of course, but more importantly it must shed a new light on a biological question of interest: in other words, it must do "better" than a non-integrative analysis.

On the biological level, a systemic vision of the functioning of the cell perfectly motivates the development of methodologies for integrating omic information. How could we actually understand the regulations of the cell without studying or understanding the numerous molecular interactions that take place therein: DNA-DNA, DNA-RNA, RNA-protein, etc. Nonetheless, omics data integration is not an easy task. It is not a miraculous solution and the demonstration that an integrative analysis provides a more complete biological picture than a non-integrative analysis is not always straightforward. We mention here very briefly some of the statistical difficulties associated with data integration (Ritchie et al. 2015).

### P.3.1.1. *Heterogeneous and complex data*

One of the first difficulties that comes across is certainly data diversity. For example:

1) data with very different formats have to be integrated: graphs, matrices, signals, etc.;

2) data corresponding to a wide range of molecular scales have to be integrated, for example, transcriptomic and proteomic data;

3) unbalanced datasets where some samples are not present in all the datasets have to be integrated.

### P.3.1.2. *Quality data*

As Ritchie et al. (2015) reminded us well, before integrating data, it is necessary to analyze each dataset separately and validate its quality. To obtain high-quality results from an integrative analysis, high-quality data are necessary.

### P.3.1.3. *High-dimensional data*

In genomics, we are often faced with the problem of high dimensionality (Giraud 2014): the number of variables $p$ (genes, proteins, transcripts) is often much larger than the number of observations $n$ (individuals, samples). Integration tends to make the problem worse. For simplicity reasons, let us assume that in each dataset $d$ to be integrated, the same $n$ samples are observed and that we measure $p_d$ variables. If in each dataset, we already have $n \ll p_d$, a fortiori $n \ll \sum_d p_d$.

One solution for mitigating the importance of this problem consists of reducing the size of each dataset. For this purpose, there are many existing techniques, for example, *data mining* techniques or even the use of knowledge bases.

## P.3.2. *Omic or multiomic knowledge integration and acquisition*

The focus is often on the need for multi-omics data integration. This need is undeniable. However, at the statistical level, we should not forget the need for mono-omic integration. A large number of classical analysis tools model biological entities independently (or almost independently). For example, for the study of RNA-seq data, differential analysis is most often used and genes are analyzed almost independently (Robinson et al. 2010; Love et al. 2014). There is a form of integration at the level of the estimation of the overdispersion parameter or even of the analyses of *pathways*. This integration already raises very important statistical difficulties. However, more should be done in modeling dependencies within a type of omics data (see Chapters 4 and 5, for example).

Clearly, integrating data should make it possible to take advantage of the very large number of datasets already available and perform powerful meta-analyses. A more data-driven, computational and simulation-based science is often predicted. Nevertheless, we think it is important to take the integration into consideration during the knowledge acquisition process (see Figure 1 of Camacho et al. (2018)). In this framework, an important question is to know how "easily integrable" data can be generated. In statistics, this is often referred to as "experimental design". The answer will obviously depend on what is to be biologically predicted or understood and which validation techniques are available.

## P.3.3. *Presentation of the second part*

In summary, omics data integration seems to be a key objective for a more integrative and systemic biology. At the statistical level, there are still some methodological obstacles remaining, in particular high-dimensionality, managing missing data, prediction within uncertain contexts and validation.

Moreover, we should not forget the objective: that is, to answer a biological question. Defining this question is not always simple. Does a biological process of interest have to be predicted or understood? Will the analysis be supervised, unsupervised or semi-supervised? What are the implicit or explicit assumptions of the analysis performed, are they consistent with the biological question?

The statistical chapters in this book hopefully illustrate how integration allows progress to be made concerning a particular biological point of interest, the diversity of methodological approaches and some of the difficulties encountered. Chapters 4 and 5 address mono-omic data integration for predicting a phenotype. Chapters 6 and 7 present exploratory techniques for multiomics analysis.

We asked the authors to present their work in a pedagogical manner and to provide the codes of their analyses and simulations. A reading committee composed of statisticians, mathematicians, bioinformaticians and biologists was able to appreciate and validate their efforts. We hope that this will make these chapters accessible to as many people as possible. We would like to thank the authors of the chapters and the reviewers for their work.

## P.4. References

Camacho, D.M., Collins, K.M., Powers, R.K., Costello, J.C., Collins, J.J. (2018). Next-generation machine learning for biological networks. *Cell*, 173(7), 1581–1592.

Chen, H., Yu, T., Chen, J.Y. (2012). Semantic Web meets integrative biology: A survey. *Briefings in Bioinformatics*, 14(1), 109–125.

Giraud, C. (2014). *Introduction to High-dimensional Statistics*. Chapman & Hall/CRC Press, London.

Love, M.I., Huber, W., Anders, S. (2014). Moderated estimation of fold change and dispersion for RNA-seq data with DESeq2. *Genome Biology*, 15(12), 550.

Ritchie, M.D., Holzinger, E.R., Li, R., Pendergrass, S.A., Kim, D. (2015). Methods of integrating data to uncover genotype–phenotype interactions. *Nature Reviews Genetics*, 16(2), 85–97.

Robinson, M.D., McCarthy, D.J., Smyth, G.K. (2010). Edger: A bioconductor package for differential expression analysis of digital gene expression data. *Bioinformatics*, 26(1), 139–140.

Zheng, C.L., Ratnakar, V., Gil, Y., McWeeney, S.K. (2015). Use of semantic workflows to enhance transparency and reproducibility in clinical omics. *Genome Medicine*, 7(73).

April 2023

# PART 1

# Knowledge Integration

# 1
# Clinical Data Warehouses

Maxime WACK[1,2] and Bastien RANCE[1,2]
[1]Hôpital Européen Georges Pompidou, AP-HP, Paris, France
[2]Centre de Recherche des Cordeliers, INSERM, Université Paris Cité, France

## 1.1. Introduction to clinical information systems and biomedical warehousing: data warehouses for what purposes?

Patient care in hospitals, private practices and any healthcare facilities produce a large amount of information: information in the form of text, such as clinical reports and prescriptions; information in the form of images, such as X-rays, scans and MRIs, as well as anatomical and pathological examinations; structured information in the form of key-value pairs, such as the results of biological laboratory examinations or codes derived from standardized terminologies (e.g. the international classification of diseases). Already useful for decision-making during care, these data can be given a new lease of life in clinical data warehouses, which allow them to be reused for research projects and, in particular, for improving care. Unlike data collected in highly standardized clinical studies, making use of selected and controlled patient profiles, health data warehouses integrate information known as "real life" data. The recorded data reflect practices and their evolutions, the use of treatments outside the planned frameworks and all other forms of incongruities

---

For a color version of all figures in this chapter, see: http://www.iste.co.uk/froidevaux/biologicaldata.zip.

*Biological Data Integration,*
coordinated by Christine FROIDEVAUX, Marie-Laure MARTIN-MAGNIETTE
and Guillem RIGAILL © ISTE Ltd 2023.

related to the modification of tools to adapt them to the requirements of daily patient care. Reusing becomes difficult to achieve due to the complexity of the data models associated with hospital information systems (HISs). Their design, oriented around information collection, patient monitoring and hospital management, is not adapted to the challenge that flexible and facilitated reuse presents.

### 1.1.1. Warehouse history

Clinical data warehouses were developed to allow the integration of heterogeneous data in a single location and according to a unified data model. Large clinical data warehouses began in the United States during the 1990s. In 1994, the Columbia University Medical Center in New York created a warehouse with the main objective of supporting clinical trials (Chelico et al. 2016). In Boston, Partners Healthcare Inc. (a company grouping together the Massachusetts General Hospital, the Brigham Hospital and the Women's Hospital) published in 2003 the description of a graphical interface named *Research Patient Data Registry* (RPDR) (Murphy et al. 2010), making possible the use of their data repository by non-computer specialists. This warehouse would evolve to become in 2004 the i2b2 solution (Murphy et al. 2010) (*Informatics for Integrating Biology and the Bedside*). The i2b2 warehouse has been widely used for several years by many institutions in the United States and around the world. In France, the Georges Pompidou European Hospital has been using the i2b2 model to develop its warehouse since 2008 (Zapletal et al. 2010). The AP-HP (Assistance publique-Hôpitaux de Paris) also deployed a common data warehouse for its 39 hospitals in 2016 using the i2b2 model at first. Further clinical data warehouse experiments and models have been developed in parallel and subsequently. For example, Vanderbilt University built a warehouse linked to a DNA bank (BioVu) (Danciu et al. 2014). More recently, it is the OMOP CDM (*Observational Medical Outcomes Partnership Common Data Model*) of the OHDSI consortium (Hripcsak et al. 2015) (*Observational Health Data Sciences and Informatics*) initially designed for use cases in pharmacovigilance, which seems to be adopted as an international reference.

### 1.1.2. Using data warehouses today

Apart from simple reuse for clinical research, data warehouses have proven their usefulness for many purposes (decision support, cohort building, predictive model validation, pharmacovigilance). Today, they are at the heart of translational medicine, which aims to translate in a practical manner the results obtained in the context of research for improving patient care. The data warehouse closes the loop of a virtuous circle by enabling the reuse of data generated in the context of medical care for the benefit of research.

## 1.2. Challenge: widely scattered data

From the patient's arrival to his or her diagnosis and possible surgery, including bioassays, drug prescriptions, nursing procedures and observations, as well as all the elements involved in keeping a medical record, a patient's stay generates a large amount of very varied information, which is now increasingly often collected in the form of computerized data.

The main challenge of integrating clinical data lies in the massive heterogeneity of these data produced within the biomedical context. The diversity of data sources (each software tool and each measure produce data of potential interest) and the increasing use of specialized systems contribute to the growing complexity of HISs. Hospitals keep patient files and have the software to manage the monitoring of hospital activity – the PMSI (the French program for the medicalization of information systems), LIMS (a laboratory software) and a PACS (*Picture Archiving and Communication System*) for managing imaging. Furthermore, there are complementary specialized information systems for advanced life support, genetic data or radiotherapy. The interconnection between these systems is not always obvious. It is not uncommon that, with the exception of common management of patient identities and medical stays, there is very little communication between software programs, making multi-system interrogation very difficult. Besides the complexity of the systems, some warehouses may even integrate multiple sources (from research databases, multiple hospital sites, etc.), making the problem even more acute (see Figure 1.1).

In addition to the diversity of information systems, there is the problem of structuring biomedical data. In the clinical context, for example, the most structured data are often the results of biological laboratory analyses (e.g. hemoglobin measured at 12.8 g/L), ICD 10-coded diagnostic codes (International Classification of Diseases, 10th Revision, for instance, code I10: essential (primary) hypertension), or still medical procedure codes coded using the CCAM in France (Common Classification of Medical Sects, for example, HHFA011: laparoscopic appendectomy). In contrast, data gathered by humans (doctors, nurses, physiotherapists, psychologists, researchers, etc.) are often collected as the result of a combination of free text without any structuring and partial structuring (the question may be labeled by interface terminologies (Rosenbloom et al. 2006), with the definition of a domain of possible answers). This is, for example, the case in the questionnaires of electronic patient records.

Consequently, each interaction of a new patient nature with the care device (new biological technique, new imaging technology, etc.) induces a new category of data, often with its own unique typology. Each new medical measurement technique is likely to generate the need for a new typology to be defined, and these typologies may be subject to further development. It is thus necessary for clinical data warehouses to be able to manage these multiple typologies, as well as to accept arbitrary types.

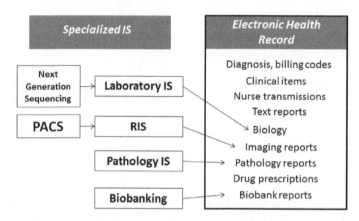

**Figure 1.1.** *Dedicated databases for a folder containing all the information*

## 1.3. Data warehouses and clinical data

### 1.3.1. *Warehouse structures*

Many software-based solutions have appeared over the years, often developed in an ad hoc manner to target the information system of a particular hospital. Later, other solutions have promoted an agnostic approach allowing the adaptation and use of information systems from other centers. Several different approaches have emerged, providing various answers to the problem of integrating heterogeneous data in a centralized manner.

#### 1.3.1.1. *Kimball model (star schema warehouse)*

Kimball's model (Kimball 1996) is oriented around observations. A central table records the facts (a biological result, a medical prescription, etc.) in a single model. A series of additional tables provide the context of the information. In the context of health data, this is often the patient, the stay, etc. The fact-based approach allows for a fairly broad definition of a patient's care, reducing it to a succession of atomic interactions with the various actors in the care process without any a priori restrictions on the number, order or type of these interactions, and without imposing any restrictions on the form that a patient's stay should take. The model is simple, very easy to implement and understandable to warehouse end users.

The main example of this approach is the aforementioned i2b2 (Informatics for Integrating Biology and the Bedside) platform, which dates back to the early 2000s. The stated objective of i2b2 was to provide an exploitation-oriented clinical data warehouse system for research, especially translational research, which facilitated linking laboratory data (biological analyses and new generation data: DNA tests, sequencing, etc.) to clinical data collected at the patient's bedside. It was therefore necessary to propose a system allowing the dynamic representation of types.

The i2b2 model follows the canonical structuring of the so-called "star schema", with a central "facts" table, surrounded by peripheral tables providing information unique to each of the keys in the central table. For example, a peripheral table will be able to contain information unique to each individual identified by a key used in the central table. A fact is the product of an atomic interaction between the patient and the hospital: a diagnosis, a bioassay result, a drug prescription, etc. For each fact observation, the whole context can be found: the patient, his or her hospitalization (stay), as well as the medical unit in which the observation was made. In the absence of a specific table for each type of data, the nature of the observation is recorded in the form of a concept (see the attribute concept_cd in Figure 1.2), a term from a more or less structured nomenclature. In i2b2, these tree-shaped nomenclatures are wrongly called "ontologies" since they do not allow for the representation of complex semantic relations. However, nomenclatures do permit the representation of taxonomies, hyponymy (*A is-a B*, for example, a lung cancer is a cancer), synonymy (*A is B*) and hyperonymy (*A contains B*) relationships. It is thus possible to express in the i2b2 formalism most of the existing medical terminologies and to propose an intuitive classification mechanism that enable the exploration of existing concepts and the creation of personalized terminologies. Each concept thus organized can be used to represent its own occurrence (a diagnosis, a procedure, etc.), or be associated with a specific value (numerical for a dosage result, a drug dosage, textual for a microbiology result, a medical questionnaire item, or even arbitrary binary information which can contain any type of document).

For example, the dosing of a natraemia of 140 mmol/L with Mr. X on November 3, 2017 during his visit to the emergency room could be coded by adding a new row to the table observation_fact containing the unique number of this particular hospitalization, the unique identifier of Mr. X employed to link his data in the HIS, the identifier of the concept of natraemia expressed in mmol/L, the identifier of the emergency department, the date and precise time of the sampling and the numerical value of the dosage (see Table 1.1).

8  Biological Data Integration

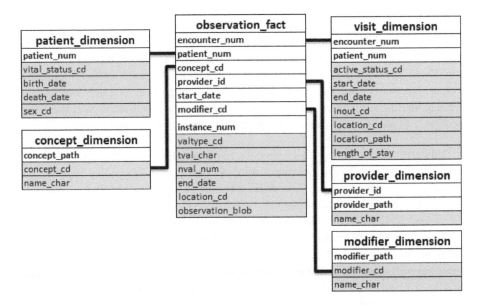

**Figure 1.2.** *i2b2 star schema*

| encounter_num | patient_num | concept_cd | provider_id | start_date | nval_num | unit_cd |
|---|---|---|---|---|---|---|
| 123456 | 112233 | BIO:NAT | STRUCT:URG | 2017-11-03 | 140 | mmol/l |
| 135691 | 112233 | BIO:NAT | STRUCT:CARDIO | 2018-10-22 | 124 | mmol/l |
| 136899 | 654332 | CIM:I10 | STRUCT:PNEUMO | 2018-05-31 | {NULL} | {NULL} |

**Table 1.1.** *Extract from the observation_fact table containing three observations for two patients*

The simplicity of Kimball's model and its ability to incorporate all sorts of data has led to the widespread adoption of i2b2. On the other hand, the Kimball model suffers from a number of drawbacks:

1) The semantics of the attributes of the fact table may change according to the type of data. For example, the record date (`start_date`) may correspond to a certain point in time (for instance, in the case of an instantaneous dosage) or to the start of an event (the start of drug administration, for instance).

2) In its original form, using only the attribute `concept_cd` to express the nature of the recorded data, the i2b2 model was not able to express the full potential of the medical data. In 2010, the designers added two additional attributes (`modify_cd` and `instance_num`) resulting in more complete information using an annotation and an occurrence number. For example, a drug such as aspirin (designated by the aspirin

concept) can be modified to express the prescribed dose. The model introduces a *modify* dose in this case. The instance is used to assign a unique identifier to the concept, thus allowing the aspirin prescription, its dose, its mode of administration, etc., to be grouped together. The use of *modify* is complex since its meaning may vary depending on the concept used.

In addition to i2b2, many warehouse models are based on the star model, or on forms derived therefrom.

### 1.3.1.2. *The Inmon model*

The first commonly accepted data warehouse model is the Inmon model (Inmon 1992). The author recommends the use of the relational model following the third normal form (3NF). This is an organization where each attribute possesses its own key, thus avoiding inconsistencies and redundancy. In 1992, this model results in maintaining a high level of information structuring, while allowing for efficient querying and use of the warehouse. Inmon proposes a *top-down* approach, the warehouse being built by analyzing existing data and following a set of principles. Tables are created based on the topics covered by the data (patients, diseases, drugs, etc.). Building these warehouses incur large costs since they require a thorough understanding of the integrated data.

The *Compton Data Model* of the OHDSI consortium (*Observational Health Data Sciences and Informatics*) is an example of the application of the Inmon model in the health domain. Each type of entity corresponds to a table (see Figure 1.3), with specific attributes (e.g. Table *Drug* – treatment – has a reason for discontinuation attribute). The tables are linked by foreign keys. Similar to the i2b2 model, OHDSI uses hierarchical nomenclatures to describe the concepts recorded (such as natremia). The OHDSI model is at the moment widely adopted by both public and private sector users.

### 1.3.1.3. *Other data warehousing models*

#### 1.3.1.3.1. From the fact to the document

The data warehouses described above are largely based on the idea that health data are at least partially structured. They are organized based on terminologies and reused. Different studies have shown that a large part of medical information, most of it in fact (Raghavan et al. 2014) is found in clinical narrative texts (medical reports, discharge letters and prescriptions) and is written in free text form. These documents not only contain medical observations, but also some of the clinicians' reasoning, assumptions and generally speaking, a detailed level of clinical observation.

To meet the need for querying and exploiting these data, "text"-oriented data warehouses have been developed (Hanauer 2006; Cuggia et al. 2011; Garcelon et al. 2018), such as Dr. Warehouse.

Dr. Warehouse is a warehouse organized around the document (examination or hospitalization report, medical observation, etc.), which replaces the structured fact of the star model. The warehouse is optimized to accelerate the query, but also the management of attributes specific to natural language, such as the negation or the subject of a property ("the patient does not exhibit such a symptom", "the patient's grandfather died of this type of illness"). Text-oriented warehouses are associated with automatic language processing methods to make the contained information queryable in a normalized and structured form.

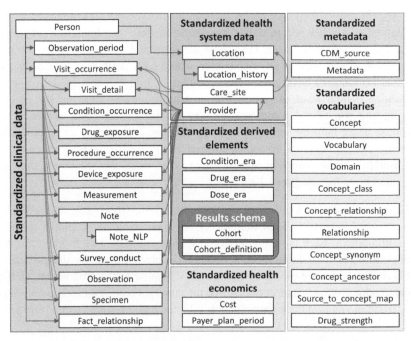

**Figure 1.3.** *OHDSI CDM data model (image distributed under Creative Commons Zero v1.0 Universal license)*

### 1.3.1.3.2. Data warehouses organized around complex representations of data

Computer research focusing on data warehouses continues and more complex models compared to the Kimball or Inmon models have been developed. These models make, for example, the management of the historicization of data, or the representation of strong semantic links (such as temporality) possible. These new warehouse models are often based on knowledge modeling (by ontological modeling, for example) and on the structure of the data sources (Khnaisser et al. 2015).

## 1.3.2. *Warehouse construction and supply*

Warehouses based on structured data representation models most often make use of Structured Query Language-based (SQL) relational database management systems These systems allow the implementation of the described data structures and the constraints necessary for their operation. Implementations based on systems known as non-SQL (NoSQL) systems (CouchDB, MongoDB, Redis, etc.) that do not use classical relational databases but key value storages which can refer to documents are beginning to emerge for clinical data warehouses, a few years after their mainstreaming in pure computer applications.

The database management system having resolved the technical dimensions, and the warehouse architecture the data representation and query structure, the main difficulty in building a warehouse appears in the integration of the data produced by the institution in the chosen warehouse solution. This is the *Extract, Transform and Load* (ETL) process (Murphy et al. 2006) (see Figure 1.4). The ETL process can include a de-identification process during the transformation stage, for example, removing the patient's personal information from the text of the reports. For each piece of data to be integrated, we must know how it can be extracted from the source information system in an automatic, programmed, regular and systematic way, and ideally be able to detect changes in the structure of the imported data, or the errors during its extraction. The data must then undergo one or more processing steps to transform it from the original format to the one accepted by the destination solution (e.g. to associate the different keys to an observation in the star model, to harmonize the expression units of the same biological result between several laboratories, to pseudonymize a hospitalization report, etc.). Finally, the data are actually loaded onto the warehouse. It is possible to choose between several data loading strategies, from the total regeneration of the warehouse in an iterative manner with a given period to the incremental data supply in real time.

## 1.3.3. *Uses*

The computerization of the various components of the HIS described in section 1.2, as well as the medical record (*electronic health record* (EHR)) over the last 20 years, has enabled collecting a large amount of digitized data relating to health care, each type in their original system and in their own format. The EHR is deployed within hospitals and is specific to the institution. The shared medical record is another computer-based initiative, over the entire French territory, which aims to enable the sharing of care data between the actors of city medicine and hospitals. At the time of writing, the majority of institutions have an EHR, but only a minority of the population and practitioners have adopted the shared medical file.

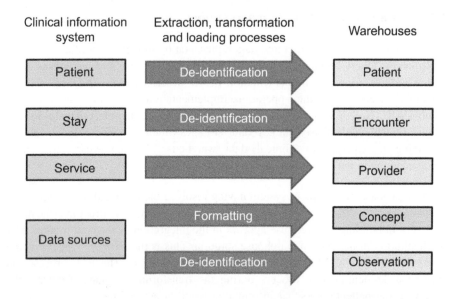

**Figure 1.4.** *Data extraction, transformation and loading*

In clinical research, the evaluation of the feasibility of a study, the constitution of cohorts, the selection of patients in retrospective studies and data collection, even data originating from health care, are costly tasks in terms of time, money and personnel. The collection of these data in a routine computerized system is therefore of great interest for their reuse within the context of clinical research, but also for the production of health indicators or for managing the institution.

HIS data are generally accessible in the context of clinical management, through various business software programs, in a so-called "vertical" manner: it is possible to access all of a patient's information from each program, or even all of a patient's information in a single EHR program in the case of the integration of data originating from business applications. The utilization itself for research or strategic orientation purposes, on the other hand, requires a "horizontal", population-based view of the data, enabling access to information for all or a subset of patients. This form is provided by clinical data warehouses whose interest for research has driven the initial developments.

1.3.3.1. *Utilization in research*

The centralized availability of clinical data from an institution, with the possibility of selecting patients according to multiple criteria (clinical, biological, pharmacological, genetic, etc.), and transversally on the units that received them, was first considered for its interest in research.

The most "trivial" use for such a tool is in the creation of cohorts (Stephen et al. 2003). Access to all care data makes it possible to search for all patients corresponding to a list of inclusion/non-inclusion criteria for a given study. It is thus possible to quickly compile a list of patients known to the institution and eligible for inclusion in a new study, where paper-based filing or data fragmentation in the HIS implied long and complex tasks. Previously, it was necessary to achieve an initial rough selection of eligible patient files against a limited number of available criteria (e.g. the diagnostic or medical procedure code produced within the framework of the French program for the medicalization of information systems (PMSI), or even the collection of data of interest over time by practitioners inside a department), and then to consult these files manually and verify each of the selection criteria. At the present time, this task is simplified with the centralization of data in a warehouse. And after all, if the researcher still has to address other difficulties derived from coding inaccuracies, from the particularities of code use linked to various factors, both epidemiological and economic, or even from the simple technical challenge of writing such a query, and that a manual validation of the inclusions is still necessary in all cases, it does not change the fact that this effort results in being nevertheless largely accelerated and the volume of files to be checked greatly reduced.

In addition to the identification of patients on a list of criteria, the presence of the data itself makes it possible to conduct studies entirely "on warehouse". In this case, these will be retrospective case–control type studies, which can be directly conducted with the data extracted following the selection of patients. For example, it will be possible to select a first list of patients presenting a particular symptom, the cases and a second list of patients not presenting this sign, but which can be compared on the basis of other data (age, sex, notable history, etc.), the controls and then to search for all drug administrations within their medical history in order to identify potential molecules responsible for this adverse effect.

The combination of the two approaches enables "retro-prospective" studies to be carried out, comprising both the selection of patients of interest, as well as the extraction of historical data from the warehouse, and prospective data, collected either by means of iterative warehouse data extraction (produced during the treatment), or by conventionally capturing research-specific data (which would not be systematically collected during the treatment). A new research project can thus eventually be initiated virtually several years in advance on account of the data already collected during the current treatment care.

Clinical data warehouses are therefore proving to be valuable tools for complementing traditional clinical research, allowing for the selection of patients, the automatic collection of a large number of variables from treatments, or even for conducting studies in a comprehensive manner.

Near real-time access to data during its production enables the implementation of systems for various vigilances, and pharmacovigilance in particular. The collection of drug prescription information combined with systematic diagnose PMSI coding and text mining (medical observations, hospitalization reports) results in scaling up pharmacovigilance and in being able to conduct phase four trials for assessing drug safety after its introduction in the market. Furthermore, in situations where routine vigilance tasks, such as monitoring nosocomial infections, require cross-referencing data from multiple sources (microbiology, biology, admissions, surgical procedures, etc.), warehouses offer a single access and detection solution for these events.

### 1.3.3.2. *Other uses*

The use of warehouses is not, however, limited to the scaling up of traditional research methods, and opens up new areas of research as well as new applications for its use in daily clinical practice.

The data available are high-dimensional data, in other words, a large number of different variables are collected for each patient. They concern a large number of patients, from a few tens of thousands for a small hospital to several million for a large hospital, over a period of several years of activity. Therefore, some establishments are in charge of data covering up to 20 years of history (Jannot et al. 2017). The extent of these data allows new types of studies to be conducted, particularly those in the field of Big Data: research on typical patient profiles, analysis of mass textual data, analysis of medical imagery, research on drug interactions, research on data quality, research on epidemiological trends in the patient pool of the institution, etc.

Clinical data warehouses have also have proven to be extremely effective in routine treatments: on the one hand, through the collection of data employed for designing decision support systems (Clinical Decision Support Systems (CDSS)) (Boussadi et al. 2012), thus creating a key virtuous loop in translational research (see Figure 1.5) (Bréant et al. 2007), but also for the evaluation of care practices in the institution, by capturing the completeness of the health care and enabling a more faithful introspection which can be conducted at a higher frequency.

These systems are mainly used for common clinical situations that generate enough data to produce statistically valid results. Nonetheless, rarer pathologies are also taken into account, since the recorded knowledge of a large number of patients can be used to search for similar clinical situations in the history of their care. For example, a doctor receiving a patient suffering from a rare disease, or having an atypical clinical presentation, will be able to try to find one or more similar cases among the patients previously treated in the establishment (Frankovich et al. 2011; Garcelon et al. 2017). He will thus be able to adapt his treatment, guided by the knowledge of previous experiences of the same nature.

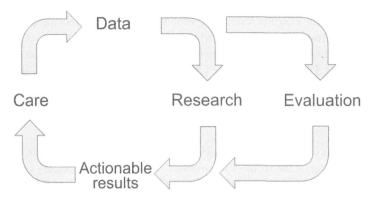

**Figure 1.5.** *Virtuous circle of translational research and practice evaluation*

In addition to their contribution to research and care, clinical data warehouses can also prove their usefulness for the management of establishments, due to the global coverage of the care activity produced by the hospital. It becomes easy to produce indicators which are updated in near real time, thus offering a privileged view of this activity. The fact that the earliest data integrated into the warehouses are the medical information from the PMSI, along its coding of medical diagnoses based on the ICD10 and procedures on the CCAM, since it is necessary for the operation of the hospital, makes the tool directly attractive for helping to develop, evaluate and achieve the full value of care strategies (e.g. the shift toward outpatient care). The medical information department (MID) teams responsible for producing PMSI data and whose quality of information coding is a prerequisite for proper valuation, see it as a support tool for automating this coding, relieving the teams of routine tasks and allowing them to devote more time to the most complex cases.

## 1.4. Warehouses and omics data: challenges

The evolution of methods in biology, especially the appearance of high-throughput methods or within the medical field, has led to the emergence of new data of very large dimensions exhaustively covering a specific domain, which are now referred to as "omic" data. Genomics, through genetic sequencing, produces billions of data points for a single patient, one per sequenced nucleotide base, even before the interpretation that can be made therefrom. Medical connected objects, such as connected pacemakers or insulin pumps, can also produce large volumes of data, with a precision reaching the millisecond. Since the goal of omics methods is the complete capture of a phenomenon, each new technique, or existing technique that result in a gain of resolution, implies generating a vast amount of non-trivial structure data. Similarly to laboratory measurement data, these data are most often accompanied by an interpretation. Is a variant clinically relevant or silent? The problem closely resembles the one deriving from biological assays such as natraemia or hematocrit, namely interpreted by comparison to reference values. Models such as i2b2 or OMOP CDM include in their schemas columns allowing a classification of biological results given in numerical form. In i2b2, a tag is used to indicate whether the result is above or below the normal values, and OMOP CDM records the reference values in its schema.

Omics data present additional challenges: different levels of granularity may be required depending on the focus of the research, and it may be necessary to represent complex relationships between data (e.g. a gene belongs to a signaling pathway). These relationships may vary depending on the nature of the data recorded and the questions asked.

### 1.4.1. *Challenges of data volumetry and structuring omic data*

In clinical data warehouses, data processing presents two major challenges: (1) volumetry and (2) the representation of the relationships existing between raw data and the different levels of interpreted data.

1) Omics data generate volumes several orders of magnitude greater than those previously stored in warehouses: a full genome sequence exposes several millions of variants, while clinical data represents at most hundreds of facts per individual. The models that warehouses commonly installed in hospitals make use of are not well suited to the manipulation and importation of such quantities of information. An alternative device could consist of assimilating the dataset (all the variants produced) to a single complex object. Such a solution properly allows for information integration and keeps the volumetry comparable with the other types of data, but makes the interrogation of the omics data dependent on the adaptation of the tools.

2) Raw omics data are analyzed with respect to references (genome version, reference variant databases, etc.). Warehouses such as i2b2 or OHDSI OMOP are not (or barely) capable of recording relationships between data (only between data and patients, or data and stay). Relationships such as "a gene belongs to a pathway", must be represented by duplication of information (one line for the variant, one line for the gene, one line for the signaling pathway). The redundancy of the information poses problems for the maintainability of the information and results in very complex queries.

## 1.4.2. *Attempted solutions*

A number of research platforms have been developed to address the specific challenges caused by omics data. They are intended to facilitate the combined interrogation and analysis of clinical and omics data (Canuel et al. 2015). In this section, we discuss two proposals: the expansion of a clinical data warehouse and a platform dedicated to omics data, but capable of integrating clinical data.

### 1.4.2.1. *i2b2 and tranSMART*

The i2b2 model has been used natively to enable storage and querying omics data. In the article by Murphy et al. (2017), the authors compare three methods for integrating omics data into an i2b2.

1) The first approach "pure i2b2" breaks down the data into "facts" and relies on a dedicated terminology, the *sequence ontology* (Eilbeck et al. 2005), adapted to the i2b2 model.

2) The second approach is based on the open-source software tranSMART. This software was initially developed for the pharmaceutical industry as a collaborative platform for preclinical studies. The software was then opened to the community as an open source application. TranSMART is designed to help scientists build and refine hypotheses by exploring correlations between phenotypes and omics data. The data structure used by tranSMART is largely based on i2b2: a star model for clinical data. Omics data are hosted in dedicated tables added to the model.

3) The third proposal links the i2b2 application to a NoSQL search engine. Clinical data are integrated in the i2b2 star model and a module added to the application graphical interface allows the graphical creation of a query to a NoSQL database (CouchDB).

In summary, the first two solutions, i2b2 and tranSMART, store information in a single relational database system, while the third solution, a hybrid one, separates clinical data from omics data. It seems that none of these three approaches has become established in warehouses, requiring either infrastructure and software redesigns or the implementation of ETLs.

### 1.4.2.2. *cBioPortal*

A widely used solution in the world of oncology is cBioPortal (Cerami et al. 2012). This is software developed by the Memorial Sloan-Kettering Cancer Center (MSKCC) in New York, USA. cBioPortal is an open-source platform designed to facilitate researchers in oncology access to datasets produced within the framework of large cancer genomics projects (e.g. *The Cancer Genome Atlas* (TCGA) (Botsis et al. 2010)). cBioPortal is dedicated to the field of oncology and integrates data such as simple phenotypic information, survival data, progression data, but especially omics data (DNA, RNA, protein, etc.). The software provides advanced visualization, analysis and data access features. There is an online version available, and the software can be freely downloaded

cBioPortal uses a "classic" relational model mostly oriented around omics data. Clinical data are limited in number and expressiveness.

## 1.5. Challenges and prospects

The technical-scientific foundations underlying clinical data warehouses have been laid and have led to successful achievements. However, the emergence of omics data has made it possible to identify their weaknesses in addressing the explosive digitization of medical data, if not all of the data pertaining to human existence, and likely to provide new relevant sources.

Researchers, such as public authorities, now have a better understanding of the implications and challenges in terms of security and ethics of this massive digitization, particularly because of the unprecedented predictive power it affords. These predictions are produced by machine learning methods, increasingly referred to as AI, whose success is entirely linked to the volume of information available, which in turn also generates information by inference, for example, by interpreting radiological images or text. The issue of the quality of incoming data as well as of the production of data, and their traceability, is thus becoming essential in situations where warehouses need to serve as real medical tools.

### 1.5.1. *Toward general-purpose warehouses*

Omics data are not the only source of complex, large-scale, voluminous data. One obvious example is that of medical imaging, which has nonetheless long been on the leading edge among medical specialties when it comes to using IT resources, and whose data are still very rarely integrated into warehouses other than in the form of a report or, at best, a diagnostic code. The very nature of the images, the volumes generated which are increasingly large as the resolution of instruments improve, and the notorious problem of their automatic annotation further complicate their integration. The "simple" integration of a reference alongside other types of data already integrated provides a solution to an initial problem of accessing images through the same patient selection channel. This patient-oriented selection crucially lays the foundation for the development of image analysis methods contextualized by the entire clinical history (Murphy et al. 2006). Nevertheless, it will be necessary to be able to integrate images and their annotations in a fluid manner; this reasoning (raw data along its interpretations) can also be applied to omic data and other forms of already existing (real-time monitoring signals, dynamic recordings, etc.) or future large-scale results. This sets a clear orientation toward generalist warehouses capable of hosting all the data produced in the hospital: pathological images, microscopic images, medical imaging, radiomics, genomics, monitoring, etc.

### 1.5.2. *Ethical dimension of the implementation and the use of warehouses*

The interest of warehouses in terms of research, public health and improved patient care is undeniable.

However, the volume of data and the level of detail of the medical data existing in warehouses make it impossible to ensure data anonymization while maintaining a sufficiently relevant level of information for secondary uses. In order to best protect patient data and comply with current regulations (Data Protection Act, GDPR), warehouse managers use three types of means.

1) Technical means: directly identifying data are removed and replaced by de-identified keys (the correspondence table being stored on another system), access to the data is logged and the data are confined to isolated systems (the user, the researcher, the data scientist can login into the system, but the data cannot leave it).

2) Organizational means: the systems rely on user profiles that give access rights to data only to properly identified individuals, and patients are informed that their data will likely be stored in warehouses and refusals to use their data for research are recorded.

3) Regulatory means: the existence of warehouses is conditional on authorizations from the CNIL (French National Commission on Informatics and Liberty), and access to data and its use, for each study, is conditional on the agreement of an ethical-regulatory committee in the case of research on data, or of a committee for individual protection in the case of research involving the human person.

### 1.5.3. *Origin and reproducibility*

The origin of the information is usually recorded in the repositories, in the form of a reference to the system from which the datum is extracted, along with metadata on the times and dates of extraction. The details of the extraction mechanisms are rarely recorded. Data transformation processes (text extraction, computation, etc.) are almost never recorded. The use of tools developed in computer science and bioinformatics (origin methods and workflow managers) could greatly increase the traceability of extractions and the reproducibility of data transformation processes.

### 1.5.4. *Data quality*

Some biomedical warehouses in France and around the world have been collecting data for over 20 years. These warehouses are not only at risk of containing erroneous data (Botsis et al. 2010), which firstly creates a problem for their reuse, but also subject to biases due to the length of the period under consideration: over the years, measurement methods may have evolved, references may have changed and populations may have changed. This is particularly true in the health sector, where public health policies, the fact of opening or closing services in hospitals, etc., can have a major influence on the data collected by hospital systems.

In a study of laboratory data collected over a period of 20 years, abrupt changes in the statistical distribution of the data have been observed (Looten et al. 2019). Upon investigation, these changes were due to various reasons, namely, change in computational methods, change in measurement devices and change in information system or laboratory. This study highlights the need to accompany the data with information about the context of their generation.

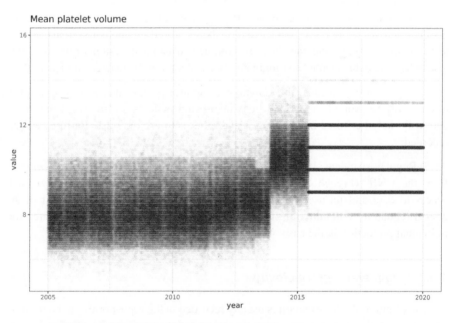

**Figure 1.6.** *Change in trend, then change in standard, and finally change in granularity of the same measure over time (since 2000)*

In the case presented in Figure 1.6, the drastic increase in the distribution mean corresponds to a change in automaton in the analysis laboratory. These changes have no consequence when care is provided to patients, since these data can be viewed along with normality limits allowing the complete interpretation of the information by the practitioner receiving the information. On the other hand, in the warehouses, context metadata are not always present, making their interpretation a posteriori dangerous, far from the context in which they were produced.

### 1.5.5. Data warehousing federation and data sharing

Despite the large number of patients and the length of follow-up in some warehouses, the need to share data remains important. By simple sharing, or ideally by a federation of warehouses, this sharing provides a means to mitigate the impact of center biases and to ensure the reproducibility of results. Large networks of federated warehouses have emerged in the United States (Weeks and Pardee 2019) and around the world. One such network was developed around the i2b2 solution, that is, SHRINE (*Shared Health Research Informatics Network*) which allows i2b2 warehouses to communicate with each other. Data does not circulate within a SHRINE network. SHRINE is limited to exchanging aggregated data corresponding to the number of patients in the different warehouses and is intended for feasibility studies (counting of patients). PCORNET (Waitman et al. 2014) (*National Patient-Centered Clinical Research Network*) is a US national patient data network that collects clinical data from health system actors (such as hospitals or patients). The data belong to major centers (the *Clinical Data Research Networks*), for example, large hospital centers, or to specific patient-led research programs called *patient-powered research networks*. The OHDSI consortium also comprises an informal network of partners. In OHDSI, there is no technical connection existing between centers, but partners can exchange queries and circulate results aggregated to scientific questions. This approach allows us to conduct studies which were previously impossible to achieve (Schuemie et al. 2018).

## 1.6. References

Botsis, T., Hartvigsen, G., Chen, F., Weng, C. (2010). Secondary use of EHR: Data quality issues and informatics opportunities. *Summit on Translational Bioinformatics*, 2010, 1–5.

Boussadi, A., Caruba, T., Zapletal, E., Sabatier, B., Durieux, P., Degoulet, P. (2012). A clinical data warehouse-based process for refining medication orders alerts. *Journal of the American Medical Informatics Association: JAMIA*, 19(5), 782–785.

Bréant, C., Borst, F., Nkoulou, R., Irion, O., Geissbuhler, A. (2007). Closing the loop: Bringing decision support clinical data at the clinician desktop. *Studies in Health Technology and Informatics*, 129(2), 890–894.

Canuel, V., Rance, B., Avillach, P., Degoulet, P., Burgun, A. (2015). Translational research platforms integrating clinical and omics data: A review of publicly available solutions. *Briefings in Bioinformatics*, 16(2), 280–290.

Cerami, E., Gao, J., Dogrusoz, U., Gross, B.E., Sumer, S.O., Aksoy, B.A., Jacobsen, A., Byrne, C.J., Heuer, M.L., Larsson, E. et al. (2012). The cBio cancer genomics portal: An open platform for exploring multidimensional cancer genomics data: Figure 1. *Cancer Discovery*, 2(5), 401–404.

Chelico, J.D., Wilcox, A.B., Vawdrey, D.K., Kuperman, G.J. (2016). Designing a clinical data warehouse architecture to support quality improvement initiatives. *AMIA Symposium*, 2016, 381–390.

Cuggia, M., Garcelon, N., Campillo-Gimenez, B., Bernicot, T., Laurent, J.-F., Garin, E., Happe, A., Duvauferrier, R. (2011). Roogle: An information retrieval engine for clinical data warehouse. *Studies in Health Technology and Informatics*, 169, 584–588.

Danciu, I., Cowan, J.D., Basford, M., Wang, X., Saip, A., Osgood, S., Shirey-Rice, J., Kirby, J., Harris, P.A. (2014). Secondary use of clinical data: The Vanderbilt approach. *Journal of Biomedical Informatics*, 52, 28–35.

Eilbeck, K., Lewis, S.E., Mungall, C.J., Yandell, M., Stein, L., Durbin, R., Ashburner, M. (2005). The sequence ontology: A tool for the unification of genome annotations. *Genome Biology*, 6(5), R44.

Frankovich, J., Longhurst, C.A., Sutherland, S.M. (2011). Evidence-based medicine in the EMR era. *The New England Journal of Medicine*, 365(19), 1758–1759.

Garcelon, N., Neuraz, A., Benoit, V., Salomon, R., Kracker, S., Suarez, F., Bahi-Buisson, N., Hadj-Rabia, S., Fischer, A., Munnich, A. et al. (2017). Finding patients using similarity measures in a rare diseases-oriented clinical data warehouse: Dr. Warehouse and the needle in the needle stack. *Journal of Biomedical Informatics*, 73, 51–61.

Garcelon, N., Neuraz, A., Salomon, R., Faour, H., Benoit, V., Delapalme, A., Munnich, A., Burgun, A., Rance, B. (2018). A clinician friendly data warehouse oriented toward narrative reports: Dr. Warehouse. *Journal of Biomedical Informatics*, 80, 52–63.

Hanauer, D.A. (2006). EMERSE: The electronic medical record search engine. *AMIA... Annual Symposium Proceedings*, 2006, 941.

Hripcsak, G., Duke, J.D., Shah, N.H., Reich, C.G., Huser, V., Schuemie, M.J., Suchard, M.A., Park, R.W., Wong, I.C.K., Rijnbeek, P.R. et al. (2015). Observational Health Data Sciences and Informatics (OHDSI): Opportunities for observational researchers. *Studies in Health Technology and Informatics*, 216, 574–578.

Inmon, W.H. (1992). *Building the Data Warehouse*. John Wiley & Sons Inc., New York.

Jannot, A.-S., Zapletal, E., Avillach, P., Mamzer, M.-F., Burgun, A., Degoulet, P. (2017). The Georges Pompidou University Hospital clinical data warehouse: An 8-years follow-up experience. *International Journal of Medical Informatics*, 102, 21–28.

Khnaisser, C., Lavoie, L., Diab, H., Ethier, J.-F. (2015). Data warehouse design methods review: Trends, challenges and future directions for the healthcare domain. In *New Trends in Databases and Information Systems*, Morzy, T., Valduriez, P., Bellatreche, L. (eds). Springer International Publishing, Cham.

Kimball, R. (1996). *The Data Warehouse Toolkit: Practical Techniques for Building Dimensional Data Warehouses*. John Wiley & Sons Inc., New York.

Looten, V., Kong Win Chang, L., Neuraz, A., Landau-Loriot, M.-A., Vedie, B., Paul, J.-L., Mauge, L., Rivet, N., Bonifati, A., Chatellier, G. et al. (2019). What can millions of laboratory test results tell us about the temporal aspect of data quality? Study of data spanning 17 years in a clinical data warehouse. *Computer Methods and Programs in Biomedicine*, 181, 104825.

Murphy, S.N., Mendis, M.E., Berkowitz, D.A., Kohane, I., Chueh, H.C. (2006). Integration of clinical and genetic data in the i2b2 architecture. *AMIA... Annual Symposium Proceedings*, 2006, 1040.

Murphy, S.N., Weber, G., Mendis, M., Gainer, V., Chueh, H.C., Churchill, S., Kohane, I. (2010). Serving the enterprise and beyond with informatics for integrating biology and the bedside (i2b2). *Journal of the American Medical Informatics Association: JAMIA*, 17(2), 124–130.

Murphy, S.N., Avillach, P., Bellazzi, R., Phillips, L., Gabetta, M., Eran, A., McDuffie, M.T., Kohane, I.S. (2017). Combining clinical and genomics queries using i2b2 – Three methods. *PLoS One*, 12(4), e0172187.

Raghavan, P., Chen, J.L., Fosler-Lussier, E., Lai, A.M. (2014). How essential are unstructured clinical narratives and information fusion to clinical trial recruitment? *AMIA Joint Summits on Translational Science Proceedings*, 2014, 218–223.

Rosenbloom, S.T., Miller, R.A., Johnson, K.B., Elkin, P.L., Brown, S.H. (2006). Interface terminologies: Facilitating direct entry of clinical data into electronic health record systems. *Journal of the American Medical Informatics Association: JAMIA*, 13(3), 277–288.

Schuemie, M.J., Ryan, P.B., Hripcsak, G., Madigan, D., Suchard, M.A. (2018). Improving reproducibility by using high-throughput observational studies with empirical calibration. *Philosophical Transactions of the Royal Society A: Mathematical, Physical and Engineering Sciences*, 376(2128), 20170356.

Stephen, R., Boxwala, A., Gertman, P. (2003). Feasibility of using a large clinical data warehouse to automate the selection of diagnostic cohorts. *AMIA... Annual Symposium Proceedings*, 2006, 1019.

Waitman, L.R., Aaronson, L.S., Nadkarni, P.M., Connolly, D.W., Campbell, J.R. (2014). The greater plains collaborative: A PCORnet clinical research data network. *Journal of the American Medical Informatics Association: JAMIA*, 21(4), 637–641.

Weeks, J. and Pardee, R. (2019). Learning to share health care data: A brief timeline of influential common data models and distributed health data networks in U.S. health care research. *eGEMs (Generating Evidence & Methods to Improve Patient Outcomes)*, 7(1), 4.

Zapletal, E., Rodon, N., Grabar, N., Degoulet, P. (2010). Methodology of integration of a clinical data warehouse with a clinical information system: The HEGP case. *Studies in Health Technology and Informatics*, 160(1), 193–197.

# 2
# Semantic Web Methods for Data Integration in Life Sciences

Olivier DAMERON
*Université de Rennes, INRIA, CNRS, IRISA, France*

The life sciences are both inherently complicated, because of the large number of different elements involved, and complex, because of the strong interdependence of these elements (Bodenreider and Stevens 2006; Bult 2006). They typically seek to identify both general rules and weak signals in a field where intra- and inter-individual variability and a gradient of situations ranging from healthy to pathological are intermingled. This situation is found replicated at different levels of granularity from molecules to organisms and ecosystems. To make matters even worse, data are generally noisy and incomplete.

Until recently, the small amount of data available in digital form and limited processing capabilities imposed the dual constraint of working with fragmented domains (precise and narrow, or broad but shallow) and making use of simplifying assumptions (Blake and Bult 2006).

The evolution of data acquisition capacities, as well as of methods and structures which allow their analysis (the Internet, computing grids, etc.), has led to an

---

For a color version of all figures in this chapter, see: http://www.iste.co.uk/froidevaux/biologicaldata.zip.

*Biological Data Integration*,
coordinated by Christine FROIDEVAUX, Marie-Laure MARTIN-MAGNIETTE and Guillem RIGAILL © ISTE Ltd 2023.

explosion in the production of data available in complementary domains, such as omics data, phenotypes, pathologies, micro- and macro-environment (Aldhous 1993; Cannata et al. 2005; Blake and Bult 2006; Bellazzi et al. 2011). This new situation suggests the possibility of moving beyond the old fragmented approach to address the complexity of the life sciences in an integrated and systematic manner.

Section 2.1 presents the characteristics of life science databases, identifies the requirements associated with their use in systematic analyses and compares these requirements to the major current approaches. Section 2.2 presents the general principles of Semantic Web technologies and shows how they could address the previously mentioned requirements. Finally, section 2.3 presents current research areas for omics data integration.

## 2.1. Data-related requirements in life sciences

Presently, we have entered an era of large-scale data production in the life sciences, and these data are available in an electronic form. However, it is clearly apparent that this is not enough to address the complexity of the living world.

The way in which these data are processed has thus become a field of study in its own right. This led Lenoir to claim as early as 1998, somewhat provocatively, that biology had become part of information sciences (and in a way, this applies to all experimental sciences). Two aspects can be considered:

– The first one obviously concerns the development of analysis methods capable of processing data presenting all the characteristics exposed at the beginning of this chapter. This is referred to as *data science*.

– The second concerns the representation of data and their relationships, as well as methods to query them such that analysis is allowed. This is called *data engineering*.

This chapter deals with the second aspect.

### 2.1.1. *Databases for the life sciences*

The BioMart portal[1] provides an interface allowing for unified access to more than 800 biological datasets hosted in some 40 databases around the world and covering in particular genomic, proteomic or cancer (Smedley et al. 2015) aspects. *Nucleic Acids Research* identifies approximately 1,600 biological reference bases[2] (Rigden and Fernández 2019). This "data deluge" (Aldhous 1993) is the name

---

1. Available at: http://www.biomart.org/.
2. Available at: http://www.oxfordjournals.org/our_journals/nar/database/cap/.

used in life sciences for the more general phenomenon of Big Data, with the particular characteristics that they involve the largest quantities of data, and that these data are highly interconnected (Stephens et al. 2015). Futhermore, this trend is likely to increase (Stephens et al. 2015). In addition to these reference databases, there are a multitude of project-specific databases and clinical data warehouses (see section 1.1).

In order to analyze the requirements related to life sciences data, it is thus necessary to take into account the large number of reference databases, at the same time resolving the contradiction between their complementary nature requiring cross-referencing, on the one hand, and the lack of interoperability of their patterns and formats, on the other hand.

### 2.1.2. Requirements

The challenge of **data integration** consists of establishing and then making the use of relationships between elements of different domains at various degrees of granularity systematic (e.g. from omic data to pathology, or vice versa for a single species or across multiple species) (Bellazzi et al. 2011). Meta-analysis of heterogeneous annotations based on pre-existing knowledge has thus enabled discoveries that would have been beyond the reach of individual analyses (Zhang et al. 2010; Rho et al. 2011). Systems biology, precision medicine and translational bioinformatics all rely on the systematic use of these relationships (Al Kawam et al. 2018).

The systematic use of the data resulting from the integration phase makes the automation of data processing even more necessary. Due to the intrinsic complexity of the life sciences, large quantities of elements and relationships representing their interdependence must thus be taken into account (Baumgartner et al. 2007; Goble and Stevens 2008).

In addition to being massive, this systematic exploitation must also address complexity (Cannata et al. 2005; Bechhofer et al. 2013). Systematic analysis of integrated data requires an element of interpretation, which relies on the knowledge of the domain (Blake and Bult 2006). This domain knowledge, also called "expertise", can be seen as the set of rules modeling under what conditions data can be used or combined to infer new data or new links between data (Levesque 2014).

The automation of systematic data analysis, based on domain knowledge, is based on the following requirements.

#### 2.1.2.1. Requirement 1: identify resources in an interoperable manner

The complementarity of the multitude of life science databases lies in the fact that several databases refer to the same entities.

In this respect, the life sciences have a long tradition of standardization that gives them an advantage over other fields (e.g. for naming people or geographical entities). Curiously, however, this is based on an uncoordinated approach where different databases each use their own identifiers. These different bases coexist and commonly refer to the identifiers of competing bases, and then mappings are performed by additional specialized bases (Côté et al. 2007). The most widely used databases end up emerging as de facto standards in their community (typically, the designation of genes is carried out between specialists of each species). They are reused by independent organizations federating the community and who are in charge of maintaining a list of unambiguous identifiers, as well as managing their obsolescence (e.g. the *Gene Ontology Consortium* for the representation of biological processes, cellular locations or molecular functions).

A similar self-organization is found for the representation of references between communities.

More recently, the requirement of implementing technical solutions to ensure interoperability between databases and to enable data analysis automation has led to more generic initiatives such as the *Life Science Identifiers* (LSIDs) (Clark et al. 2004), identifiers.org (Juty et al. 2012; Wimalaratne et al. 2015), or the research by McMurry et al. (2017) and Pierce et al. (2019).

### 2.1.2.2. Requirement 2: describe resources

Once entities are identified, databases are used to represent their description in a structured way (as opposed to free text descriptions).

The description of an entity covers three axes:

– Its **characteristics** that are its intrisic properties: for example, for a gene, its usual name, its synonyms, its start and end coordinates.

– Its **relationships** with different entities designated by their identifier: for example, for a gene, the chromosome and the strand on which it is found, the taxon to which it relates, the transcripts that it can produce. It can be seen that the description of an entity typically involves several different relationships, pointing to other entities in the same database or in external databases (in the latter case, this is referred to as *entity matching*).

– The **classes** (also called "categories") to which it belongs. These classes generally depend on the values of some of the relationships mentioned above. For example, a gene may belong to the class of protein-coding (therefore, depending on the nature of its transcripts, as opposed to the class of genes coding for transfer RNAs) and/or to the class of homeotic genes (which regulate the development of anatomical structures, so here in this case, depending on the function of the gene transcripts).

Unlike the previous point, only the instantiation relation is involved here, which no longer points to other entities, but to classes of entities from the knowledge bases.

The description of an entity thus forms a graph. Because entities may have relationships with each other, or belong to the same classes, the descriptions of a dataset often form a **connected graph**.

The description of an entity through its relationships with other entities and classes thus refers to elements of external databases and knowledge bases (requirement 1).

### 2.1.2.3. *Requirement 3: combine split descriptions*

Because of the complexity of the life sciences, the description of an entity typically involves numerous parameters. It is therefore uncommon for an exhaustive description of an entity to be available in a single database. Requirement 2 indicated that the description of an entity may reference elements in other databases. Requirement 3 indicates that this description is potentially fragmented into different databases itself.

The requirement to combine partial descriptions of an entity originating from several sources is thus derived from requirements 1 and 2.

### 2.1.2.4. *Requirement 4: query these descriptions*

The first three requirements, respectively, concerned the identification of resources and the representation of their descriptions. However, having access to these descriptions is not enough, and it is necessary to be able to query them automatically in order to perform exhaustive analyses.

It is thus necessary to have a query language that can be applied to descriptions located in one database (requirement 2), as well as to those resulting from the integration of several databases (requirement 3).

### 2.1.2.5. *Requirement 5: reason about these descriptions based on knowledge*

Finally, the knowledge of the domain is implied both in the query mechanism itself (requirement 4) *by way of* the classes of the entities and the relationships between these classes (requirement 2), and in the analysis of the result of the queries (Bechhofer et al. 2013). On these two points, Stevens et al. (2000) observe that:

> Complex biological data stored in databases often require the addition of knowledge for specifying and constraining values. [...] Much of biology is based on applying prior knowledge to an unknown entity[3].

---

3. Author's translation.

Here again, it is necessary to be able to take this into account in automated processing for performing exhaustive analyses.

This symbolic domain knowledge is formalized in knowledge bases in the form of **ontologies**, which are formal representations of knowledge in which essential classes are combined by rules that describe their structure and relationships to each other (Bard and Rhee 2004). The life sciences have a long history of modeling knowledge in the form of ontologies (Cimino 1998, 2005; Smith 2005) and their utilization for annotating data (Stevens et al. 2000; Bard and Rhee 2004; Blake and Bult 2006; Bodenreider and Stevens 2006). These ontologies are therefore already available (Stevens et al. 2000; Cimino and Zhu 2006), and their (re)use is facilitated by portals such as BioPortal (Noy et al. 2009; Whetzel et al. 2011).

From the knowledge in addition to the data descriptions, **reasoning** can be defined as a set of methods guiding the automatic traversal of the data graph or its enriching. This reasoning mechanism makes it possible to traverse relationships between two entities, between an entity and a class, and between two classes, spread over several databases. The automatic navigation rules are conditioned by the degree of ontology formalization, which ranges from simple hierarchies to semantically rich organizations (Cimino and Zhu 2006).

As for requirement 4, it is necessary to have a reasoning mechanism that applies both to descriptions and ontologies located in one database (requirement 2) and to those resulting from the integration of several databases (requirement 3).

### 2.1.3. *Common approaches: InterMine and BioMart*

Presently, for the life sciences, the main data and knowledge bases do exist, but they are distributed over silos that do not allow much interoperability.

In recent decades, this observation led to solutions specific to the life sciences to enable unified access to the various databases. InterMine is based on a centralized approach where data from different databases are integrated into a single warehouse with a predefined format (Kalderimis et al. 2014). There are about 30 such warehouses, organized by species[4]. BioMart relies on a decentralized approach through a federation of about 40 databases[5] (Smedley et al. 2015).

It should be noted that the success of these initiatives has been limited, due in part to the difficulties of implementing them centered around heterogeneous databases

---

4. Available at: http://registry.intermine.org/.
5. Available at: http://www.biomart.org/community.html.

that evolve rapidly and independently of each other. These solutions are specific to life sciences. They group together a few dozen databases and it is clear that none of them allow for scaling up to the 1,600 reference databases. Moreover, while these approaches aim to address integration requirements (requirements 1–4), they provide little in terms of knowledge-based reasoning capabilities (requirement 5).

In light of these limitations, making biological data reusable and interoperable has proven to be a major challenge (Goodman et al. 2014). Given that many other fields have also experienced a data explosion, the question must be raised whether the problem of integrating biological data is larger than life sciences themselves and should be addressed in a comprehensive and generic manner.

## 2.2. Semantic Web

The life sciences data integration and analysis initiatives that we saw in section 2.1 have developed in parallel with the emergence of the Semantic Web in the computing community in the early 2000s (Berners-Lee et al. 2001; Ruttenberg et al. 2007; Schulz et al. 2013).

The Semantic Web is an extension of the classic web that emphasizes data representation in a format that makes their meaning explicit and enables their automatic processing by associating them with ontologies (Shadbolt et al. 2006; Berners-Lee et al. 2007). This format must allow a fine data representation, their integration and their interpretation. The principles are:

– **To refine the granularity of the information**: although the classic web is organized around documents, the Semantic Web is centered on the data elements contained in these documents by identifying each one by a specific URI[6] (now an IRI[7]).

– **To explicitly represent relationships** by also identifying them by URIs: while one of the keys to the success of the classic Web is the representation of untyped links between documents (via the href in HTML), the Semantic Web relies on typed relationships (themselves identified by their URI) between data.

– **To enable the generalization and taking into account** the knowledge of the domain by means of special relationships: **instantiation** between an atomic data element (which is anecdotal) and a generality (which is universal), and the **subsumption** between two generalities. For example, instantiation relationships are

---

6. URI: Uniform Resource Identifier.
7. IRI: Internationalized Resource Identifier, which is an extension of URIs to non-ASCII characters, allowing for diacritical marks or non-Western alphabets.

used to indicate that a patient's pathology with all its specificities is an element of the set "Alzheimer's disease" (which is identified by DOID:10652 in the *Disease Ontology*). The subsumption relationship can be used to indicate that cases of Alzheimer's disease (DOID: 10652) form a subset of neurodegenerative diseases (DOID:1289).

### 2.2.1. *Techniques*

In the context of the Semantic Web, the W3C has proposed several recommendations (which are de facto standards) related to data and knowledge representation (RDF, RDFS and OWL), integration and analysis (SPARQL and OWL).

#### 2.2.1.1. *URI (and IRI) to identify the data*

To address the first requirement to identify data, the Semantic Web relies on URIs, which are an extension of URLs[8], whose syntax they preserve. For example, "http://purl.uniprot.org/uniprot/P35558" is the URI of the human protein PCKGC in the UniProt database.

URLs are therefore special cases of URIs and it is easy to see that giving the address of a document is a good way of identifying it (it is the document that is at this address). Nevertheless, this mechanism is poorly adapted to designate data inside a document (except to use the local part of a URL, after the #), to represent the relationships between these data (here the local part of the URL is not enough) or to manage fragmented data descriptions in several documents. Conversely, not all URIs are necessarily URLs and do not necessarily designate a document accessible on the Web. While it is not required, it may be convenient to have the URL for a URI that refers to the address of the description of the entity in question. This mechanism is called **URI dereferencing**. This description can even take different forms using the HTTP protocol content negotiation mechanism: typing "http://purl.uniprot.org/uniprot/P35558" in a browser loads an HTML page with a description of PCKGC intended for human users. The same address with the *mime type* application/rdf+xml returns a PCKGC description in RDF format (the same that can be obtained at the URL: http://purl.uniprot.org/uniprot/P35558.rdf).

Users can create URIs as they please, as long as they are syntactically valid, without needing to own the domain name or deploy a site. This makes URIs very easy to use. However, just because a user has the ability to create the URI "http://purl.uniprot.org/uniprot/B12345" does not necessarily mean it is appropriate: (1) adding an

---

8. URL: Uniform Resource Locator, represents the address of a document on the Web.

entity to the UniProt database is not enough, (2) UniProt administrators (or anyone else, for that matter) could very well decide to create the same URI to designate another entity, which would then lead to ambiguity and (3) by simply creating a URI such as "http://www.univ-rennes1.fr/odameron/B12345" the user would have avoided the two previous problems. While the ability to easily create URIs helps to avoid ambiguity, interoperability should be promoted by reusing existing URIs to avoid having multiple URIs that refer to the same entity. In "Cool URIs don't change"[9] in 1998, Tim Berners-Lee had begun to address good practices related to URIs (fortunately, the URL has not changed since), and this document was then extended in a W3C note "Cool URIs for the Semantic Web"[10] 10 years later.

In the remainder of this chapter, and in line with common practice, a **resource** designates any entity that can be identified by a URI:

– real entities ("http://www.wikidata.org/wiki/Q42" refers to Douglas Adams);

– groups of entities ("http://purl.uniprot.org/uniprot/P35558" refers to the set of human PCKGC proteins), including imaginary ones ("http://www.wikidata.org/wiki/Q7246" refers to the set of unicorns);

– intellectual constructs (" http://purl.obolibrary.org/obo/DOID_1289" for neurodegenerative diseases).

Finally, URIs are well suited to automatic processing but are difficult for humans to manipulate. The CURIE (*Compact URI*) are abbreviated versions comprising a prefix and a local identifier. For example, uniprot:P35558 is the shortened version of "http://purl.uniprot.org/uniprot/P35558" after associating the value "http://purl.uniprot.org/uniprot/" with the prefix uniprot. This is the mechanism that was involved when DOID:10652 was used to identify Alzheimer's disease. The user is free to choose the name of the prefixes, since these are then systematically replaced by their value to transform CURIEs into complete URIs. For further clarity, however, the same prefixes are often used, and the "http://prefix.cc" site maintains a list of frequent prefixes and their associated URI fragments.

### 2.2.1.2. *RDF to describe the data*

RDF[11] (*Resource Description Framework*) is a formalism for representing the elements describing an entity as a set of **triplets**:

– the **subject** is the first element of the triple. It is the URI of the resource that is described;

---

9. Available at: https://www.w3.org/Provider/Style/URI.html.
10. Available at: https://www.w3.org/TR/cooluris/.
11. Available at: http://www.w3.org/RDF/.

– the **predicate** is the second element of the triple. It is the URI of the relationship that is used to describe the subject;

– the **object** is the third element of the triple. It is one of the possible predicate values for the subject. This value can be a URI if the predicate describes a relationship between two resources, or a string of characters representing a simple string (e.g. for the name of a protein), a number (e.g. for the start position of a gene), a date, a Boolean value, etc. If a predicate has several values (e.g. a transcription factor that regulates several genes), as many triples must be used as there are values. This thus allows several databases to complement each other easily, as long as they use the same URIs to designate the same resources.

RDF provides a special predicate (rdf:type) to represent the instantiation relationship between a resource and a class. The predicate rdfs:label is also often found to describe a resource by a readable string.

A triple can be visualized as an arc (the predicate) going from the subject to the object. The triples of a *dataset* thus form a directed graph, consisting of related components when two triples share the same subject, the same object or when the object of one is the subject of the other. Figure 2.1 shows a simplified view of the description of the Reactome chemical reaction BR5566 that converts oxaloacetate into phosphoenolpyruvate. It is clearly visible in particular that it is an instance of the BiochemicalReaction class that comes from the BioPAX ontology, that it is associated with a string by the relation displayName, that it is part of the metabolic pathway of neoglucogenesis and that some relationships can have several values.

It consumes (via the left relationship) two molecules, one of which is oxaloacetate (SM1312), and produces (via the right relationship) three, including phosphoenolpyruvate (SM1316). Finally, it should be noted that the elements appearing in the description of this chemical reaction refer to resources from other datasets represented here by different colors, which illustrates the integration capacity of RDF. For example, taxon:9606 denotes *Homo sapiens* in the NCBI *taxonomy of species*, the PCKGC enzyme that catalyzes the reaction is the protein P35558 in UniProt, and the molecules consumed and produced by the reaction are associated with resources in the ChEBI ontology.

There are several file formats to represent RDF data: *N-triples* (one triple per line, URIs are given in full), Turtle (*N-triples extension* making it possible to use CURIE and comments), XML-RDF for a serialization in XML and JSON-LD for a serialization in JSON, or RDFa to include RDF in HTML pages. Finally, since processing massive amounts of data is one of the purposes of the Semantic Web, it is also possible to overcome the limitations of files by storing RDF data in *triplestores*, which are the RDF equivalent of databases. In addition to storing RDF data, *triplestores* can be queried using the SPARQL protocol and language by way of *endpoints* (see section 2.2.1.4).

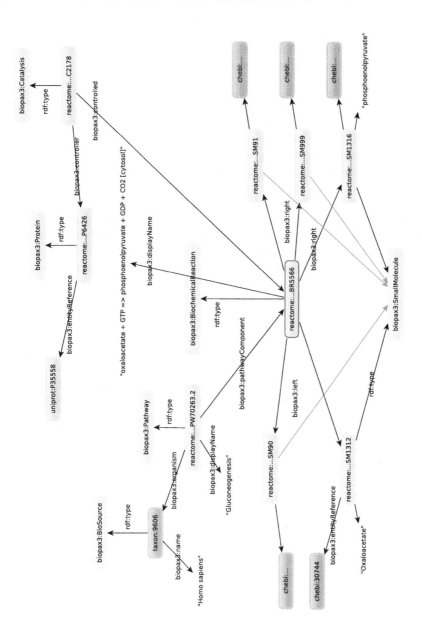

**Figure 2.1.** *Simplified view of the RDF triples describing the resource BR5566 (circled in red) of the Reactome database*

COMMENTS ON FIGURE 2.1.– *Resources are represented by rectangles and literals by strings. The predicates are the arcs, connecting the subject to the object of each triple. Some* rdf:type *relationships are grayed out for ease of reading. The resources of the Reactome database refer to resources from other databases such as the NCBI species hierarchy (in pink), proteins from UniProt (in yellow), molecules from ChEBI (in purple) or for class instantiation from the BioPAX ontology (in blue).*

### 2.2.1.3. RDFS and OWL to describe knowledge

While RDF is suitable for describing entities, their type(s) and relationships between entities, RDFS[12] (RDF Schema) and OWL[13] (*Ontology Web Language*) are models for describing the classes or predicates that form ontologies. RDFS and OWL are based on set semantics.

RDFS provides two particular resources: rdfs:Class and rdfs:Property, and two predicates: rdfs:subClassOf and rdfs:subPropertyOf. rdfs:Class is used to indicate that a resource (e.g. taxon:9606) is a set of instances (this is the set of all *Homo sapiens*). rdfs:subClassOf allows the description of the inclusion of the subclass in its superclass (e.g. here that all *Homo sapiens* are also elements of the genus *Homo*). It corresponds to the subsumption relationship that we have already seen. The principle is the same for predicates with rdfs:Property and rdfs:subPropertyOf. Ontologies in RDFS are therefore *taxonomies*, in other words, sets of classes linked by the subsumption relationships. Figure 2.1 presents the description of a chemical reaction by referring to entities of the NCBI ontologies *Taxonomy of species* (in pink) and ChEBI (in purple). Figure 2.3 reuses this description and shows how these two ontologies make the subsumption relationships between these and their superclasses explicit. The figure also shows that RDF and RDFS combine naturally, making data and knowledge integration possible.

OWL provides two resources: owl:Class and owl:Property, which are subclasses (via rdfs:subClassOf) of rdfs:Class and rdfs:Property. OWL then makes it possible to indicate whether two classes are disjoint or equivalent, or to define new classes by extension (namely, by enumerating its members exhaustively): the class of items related to at least one instance of a class by a predicate (e.g. the class of cheese pizzas is the set of pizzas in which at least one ingredient is an instance of cheese), or the class of items in which all values of a predicate are

---

12. Available at: http://www.w3.org/TR/rdf-schema/.
13. Available at: http://www.w3.org/2001/sw/wiki/OWL.

instances of a class (e.g. the class of vegetarian pizzas). Ontologies in OWL are thus much richer than the taxonomies in RDFS.

```
# prefix declarations for CURIEs
@prefix rdf:       <http://www.w3.org/1999/02/22-rdf-syntax-ns#>    .
@prefix reactome:  <http://www.reactome.org/biopax/61/48887#>       .
@prefix biopax3:   <http://www.biopax.org/release/biopax-level3.owl#>  .
# ...

# RDF triples
reactome: BR5566  rdf:type  biopax3: BiochemicalReaction   .
reactome: BR5566  biopax3: left  reactome: SM90  .
reactome: BR5566  biopax3: left  reactome: SM1312  .
reactome: BR5566  biopax3: right  reactome: SM91  .
# ...

reactome: SM1312  rdf:type  biopax3: SmallMolecule  .
reactome: SM1312  biopax3: name  "Oxaloacetate"   .
# ...
```

**Figure 2.2.** *RDF Turtle syntax representation of part of the triples in Figure 2.1. Each triple is represented on a line in the order subject, predicate, object and corresponds to one of the arcs in Figure 2.1. There is no particular order between the triples*

Figure 2.4 shows part of the OWL representation of the BioPAX BiochemicalReaction class. On the one hand, it can be seen that for the instances of this class, all the values of the `participant` relationships (especially the `left` and `right` relationships, see Figures 2.1 and 2.3) must be direct or indirect instances of the class `PhysicalEntity` and, on the other hand, `BiochemicalReaction` is disjoint from the `Degradation` and `ComplexAssembly` classes (but not from `Transportation`, for example, which explains that some conversion reactions can be biochemical transport reactions). An OWL reasoner will infer from this ontology that if a reaction is an instance of `BiochemicalReaction`, it cannot be an instance of `Degradation` or `ComplexAssembly`, and all values of the `participant` relationships must be instances of `PhysicalEntity`. Conversely, if an instance of `BiochemicalReaction` violates one of these constraints, an inconsistency will be detected by the reasoner.

RDFS allows taxonomies to be represented from inclusions of subclasses in their superclasses. OWL leads to finer descriptions and therefore richer reasoning from union, intersection and set complements, existential or universal quantifiers, and necessary and sufficient definitions. RDFS is well suited to simple reasoning about large knowledge bases, while more complex reasoning can be applied to OWL,

# 38 Biological Data Integration

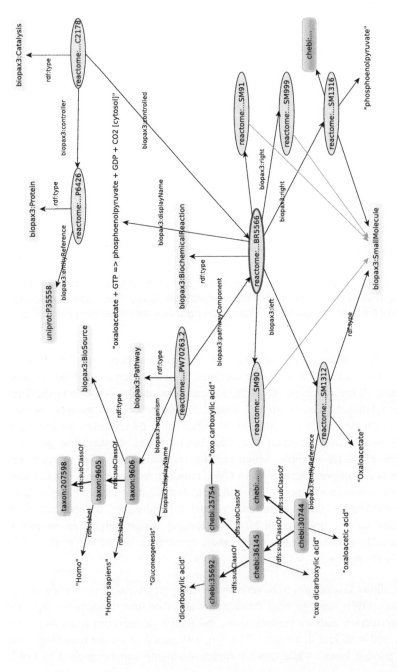

**Figure 2.3.** *Knowledge associated with the description of the chemical reaction BR5566 circled in red*

COMMENTS ON FIGURE 2.3.– *Instances are represented by ellipses and classes by rectangles. The* `rdf:type` *relationships between the rectangles and* `owl:Class` *are not represented in order not to overload the figure. Subsumption relationships* `rdfs:subClassOf` *are highlighted in bold. The knowledge obtained from ChEBI is in purple and that from NCBI Taxonomy of species in pink. It should be noted that the transitivity of* `rdfs:subClassOf` *allows generalization in both ontologies and that multiple inheritance is allowed (here in ChEBI).*

possibly incurring longer computation times. The OWL formalism consists of entities and special predicates represented in RDF (similarly to the use of `rdf:type` in RDF for designating instantiation, or `rdfs:subClassOf` in RDFS for designating subsumption). Therefore, any RDFS or OWL assertion is also a valid RDF assertion. OWL is itself an RDFS-based extension, which makes sure that any assertion in OWL is also valid in RDFS. It is therefore possible to use RDFS tools with OWL ontologies, even if they will not be able to make use of the OWL specific elements.

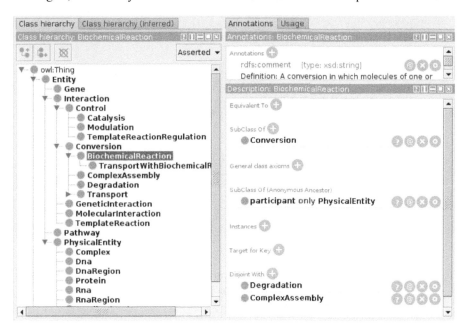

**Figure 2.4.** *OWL representation of the BioPAX* `BiochemicalReaction` *class, viewed with the Protégé ontology editor*

COMMENTS ON FIGURE 2.4.– *The left-hand side shows the class hierarchy. The right-hand part shows the constraints associated with the selected class. Here it can be seen that* `BiochemicalReaction` *is a subclass of* `Conversion`, *that it is disjoint from*

*Degradation* and *ComplexAssembly*, and that all the values of the *Participant* relationship (or of its sub-relations) must be instances of the *PhysicalEntity* class.

### 2.2.1.4. SPARQL to query

Querying data in RDF (and therefore also knowledge in RDFS and OWL, since these formats are based on RDF) is possible with the SPARQL language[14] (Pérez et al. 2009). It should be noted that SPARQL 1.1 supports most of the RDFS semantics.

The SPARQL query in Figure 2.5 enables finding in Reactome all the chemical reactions consuming oxaloacetate in humans (see Figure 2.1). Reasoning can also be performed based on the subsumption relationship by exploiting the knowledge contained in the ontologies. Query 2.6 thus generalizes the one in Figure 2.5 by finding all the chemical reactions consuming an oxo-carboxylic acid or one of its sub-classes, directly or indirectly (see Figure 2.3).

```
PREFIX   rdf : <http://www.w3.org/1999/02/22-rdf-syntax-ns#>
PREFIX   rdfs : <http://www.w3.org/2000/01/rdf-schema#>
PREFIX   reactome : <http://www.reactome.org/biopax/61/48887#>
PREFIX   biopax3 : <http://www.biopax.org/release/biopax-level3.owl#>
PREFIX   taxon : <http://identifiers.org/taxonomy/>

SELECT  ?reaction  ?reactionName
FROM  <http://rdf.ebi.ac.uk/dataset/reactome>
WHERE {

   ?reaction  rdf:type  biopax3:BiochemicalReaction  .
   OPTIONAL  { ?reaction  biopax3:displayName  ?reactionName  . }
   ?reaction  biopax3:left  reactome:SmallMolecule1312  . # oxaloacetate

   ?pathway  biopax3:pathwayComponent   ?reaction  .
   ?pathway  biopax3:organism   taxon:9606  .
}
```

**Figure 2.5.** *SPARQL query to find in Reactome all chemical reactions consuming oxaloacetate in humans (see Figure 2.1)*

COMMENTS ON FIGURE 2.5.– *In SPARQL, variable names begin with a question mark. The SPARQL engine determines the combinations of variable values that satisfy the constraints of the query from the dataset. The EBI endpoint can be used on "https://www.ebi.ac.uk/rdf/services/sparql".*

---

14. Available at: http://www.w3.org/TR/sparql11-overview/.

It is possible to query RDF files in SPARQL (by means of the Jena project[15]), as well as data contained in *triplestores* via their *endpoints* (*triplestore* refers to storage and indexing aspects, while *endpoint* refers to querying data stored in a *triplestore*). The latter use indexes to ensure good performance, even on large volumes of data. The main software programs used for deploying an *endpoint* are Virtuoso[16], Fuseki[17], RDF4J[18] and Corese (Corby et al. 2004).

```
PREFIX   rdf:  <http://www.w3.org/1999/02/22-rdf-syntax-ns#>
PREFIX   rdfs: <http://www.w3.org/2000/01/rdf-schema#>
PREFIX   reactome : <http://www.reactome.org/biopax/61/48887#>
PREFIX   biopax3 : <http://www.biopax.org/release/biopax-level3.owl#>
PREFIX   taxon : <http://identifiers.org/taxonomy/>
PREFIX   chebi: <http://purl.obolibrary.org/obo/CHEBI_>

SELECT ?reaction  ?reactionName
FROM  <http://rdf.ebi.ac.uk/dataset/reactome>
WHERE  {

    ?reaction  rdf:type  biopax3 :BiochemicalReaction  .
    OPTIONAL  { ?reaction  biopax3 :displayName  ?reactionName  . }
    ?reaction  biopax3 :left  ?reactant  .

    ?reactant  biopax3 :entityReference  ?reactantChEBI  .
    ?reactantChEBI   rdfs:subClassOf * chebi :25754 . # oxo carboxylic acid

    ?pathway  biopax3 :pathwayComponent   ?reaction  .
    ?pathway  biopax3 :organism  taxon :9606 .
}
```

**Figure 2.6.** *SPARQL query that returns in Reactome all chemical reactions consuming an oxo-carboxylic acid or one of its direct or indirect subclasses in humans*

COMMENTS ON FIGURE 2.6.– *The reasoning consists of following the* `rdfs:subClassOf` *relationship zero, one or more times from* `?reactantChEBI`*. It is represented by applying the Kleene star in the* `rdfs:subClassOf` *relationship line 17. The class hierarchy around oxo-carboxylic acid is based on the ChEBI ontology (see Figure 2.3).*

---

15. Available at: https://jena.apache.org/.
16. Available at: https://virtuoso.openlinksw.com/.
17. Available at: https://jena.apache.org/documentation/fuseki2/index.html.
18. Available at: https://rdf4j.eclipse.org/.

## 2.2.1.5. OWL reasoners for using knowledge

OWL does not have a query language, but it does not really need one because reasoning consists mainly of determining whether an entity is an instance of a class, or whether a class is a subclass of another class. The main reasoners are HermiT[19], Pellet[20] and Racer[21].

It should be noted that although most bio-ontologies are represented in OWL, few of them use its expressiveness. Most of them are actually RDFS taxonomies masqueraded as OWL (which is possible since all OWL classes are also RDFS classes), although work has shown that they would benefit from the additional primitives of OWL (Aranguren et al. 2007; Stevens et al. 2007; Hill et al. 2013).

Moreover, SWRL[22] (*Semantic Web Rule Language*) makes it possible to represent inference rules containing variables (OWL does not have variables).

### 2.2.2. Implementation

#### 2.2.2.1. bio2rdf, BioPortal and available resources

The life sciences are a prime application area for Semantic Web technologies (Post et al. 2007; Cannata et al. 2008; Bellazzi 2014), and several teams are working on life sciences data analysis using the Semantic Web, in particular within the W3C interest group *Semantic Web Health Care and Life Sciences* (HCLSIG[23]).

Major databases and knowledge bases such as UniProt[24] or the EBI[25] and NCBI (Anguita et al. 2013) databases are adopting Semantic Web technologies and are accessible via endpoints. Similarly, BioPortal (Noy et al. 2009; Whetzel et al. 2011), which lists the main ontologies in the life sciences, makes them accessible via SPARQL (Salvadores et al. 2012). Finally, the Bio2RDF project[26] promotes simple conventions for converting and integrating databases that have not yet begun their transition to RDF (Belleau et al. 2008; Cheung et al. 2009; Callahan et al. 2013).

Semantic Web technologies are now an integral part of personalized medicine and translational bioinformatics (Bellazzi et al. 2011; Chen et al. 2012). Several works

---

19. Available at: http://www.hermit-reasoner.com/.
20. Available at: https://github.com/stardog-union/pellet.
21. Available at: https://github.com/ha-mo-we/Racer.
22. Available at: http://www.w3.org/Submission/SWRL/.
23. Available at: http://www.w3.org/wiki/HCLSIG.
24. Available at: https://sparql.uniprot.org/.
25. Available at: https://www.ebi.ac.uk/rdf/.
26. Available at: https://github.com/bio2rdf.

have shown that these technologies can be used to integrate and query genotype and phenotype information (Sahoo et al. 2007; Taboada et al. 2012). In addition, Holford et al. (2012) proposed a Semantic Web-based architecture for integrating cancer-related omics data and biological knowledge.

#### 2.2.2.2. *LOD*

The **Linked Data** initiative (Bizer et al. 2009) and particularly the *Linked Open Data* project focus on the integration of data sources into Semantic Web formats. The well-known cloud of linked data[27] clearly shows the major role of life sciences through both the number of available databases and the density of their cross-references. In recent years, Semantic Web technologies have proven their relevance to the data integration requirement (Hill et al. 2013; Livingston et al. 2015).

In addition, mapping between identifiers from different databases is also facilitated by initiatives such as "identifiers.org" (Wimalaratne et al. 2015), which provides a registry associating a pair (database, identifier) in the database with a URI. Finally, once these databases are integrated, it becomes possible to query them (relatively) easily using **federated queries**, which treat these different databases as if they were a single virtual graph (Cheung et al. 2009). This point is presented in more detail in section 2.3.3.

### 2.3. Perspectives

Semantic Web technologies are a relevant solution to the requirements identified in section 2.1. They are currently being successfully used to address the complexity and heterogeneity of data in the life sciences by enabling the integration of reference databases and knowledge bases.

Nevertheless, the new perspectives that they allow raise new challenges.

#### 2.3.1. *Facilitating appropriation to users*

Although Semantic Web technologies are commonly used by data scientists, their adoption by biologists and physicians remains limited. The main factors are technical and psychological. The technical dimension is related to the difficulty of learning languages such as RDF and SPARQL, and the difficulty of implementing them by writing scripts to convert the raw data, or by deploying endpoints. The psychological aspect is related to the ease of use of spreadsheets, and the limited but intuitive processing and analysis techniques.

---

27. Available at: https://lod-cloud.net/.

These two factors explain why end-users feel dispossessed of their data once they are converted to RDF: they lose direct access to their data and must necessarily go through a *data scientist*. It is unfortunate that we are currently stuck between continuing to use spreadsheets that do not meet integration requirements, or relying on Semantic Web technologies that make the exploratory phase of data analysis more complicated and rigid by people whose expertise lies in what that data represents.

To facilitate the appropriation of Semantic Web technologies by users beyond data scientists, two levers are required:

– To facilitate the integration of their project data into the cloud of reference databases and knowledge bases. This integration must be bi-directional: on the one hand, to allow these reference bases to be exploited when analyzing data for a given project, and, conversely, to incorporate project data into the linked data cloud to foster reuse (even if not necessarily in the form of open data). Initiatives such as Datalift (Scharffe et al. 2012) are a step forward in this direction. Datalift is a platform that takes input data in different formats such as CSV or XML and makes them available by integrating them with linked data.

– To facilitate the creation of queries (possibly) concerning the user's data and the data of reference bases. Several works have already been carried out in this sense to synthesize the structure of a *dataset* (van Dam et al. 2015) or around the creation of queries such as SPARQLassist (McCarthy et al. 2012), SPARKLIS (Ferré 2014), AskOmics[28] or SPARQLbuilder (Yamaguchi et al. 2014), but none of them have established themselves as an ideal solution yet.

### 2.3.2. *Facilitating the appropriation by software programs: FAIR data*

One of the two challenges of the appropriation of Semantic Web technologies by users (section 2.3.1) is to allow them to combine their own datasets with reference databases. It is clear that this leads to a problem of scaling up. The FAIR initiative aims to address this pitfall by making data *Findable*, *Accessible*, *Interoperable* and *Reusable* (Wilkinson et al. 2016). This approach also includes data analysis workflows (see section 3.6).

The FAIR approach has shown its relevance in the field of life sciences (Rodríguez-Iglesias et al. 2016; Brandizi et al. 2018; Sima et al. 2019). It should be noted that this is supported by efforts to enhance the production of reusable datasets (Pierce et al. 2019).

---

28. Available at: https://askomics.org.

### 2.3.3. Federated queries

RDF makes it possible to take advantage of cross-references between databases (by virtue of the *entity matching* mechanism mentioned in requirement 2, section 2.1.2.2) without having to merge these databases into a single *dataset*. Several "datasets" are thus available to us in the form of *endpoints* and thus refer to each other, according to the principle of linked data. It is of course possible to separately query each endpoint, but a question may also require combining information from several endpoints (see requirement 4, section 2.1.2.4).

```
1   PREFIX   rdf :<http://www.w3.org/1999/02/22-rdf-syntax-ns#>
2   PREFIX   rdfs :<http://www.w3.org/2000/01/rdf-schema#>
3   PREFIX   biopax3 : <http://www.biopax.org/release/biopax-level3.owl#>
4
5   PREFIX   chebi : <http://purl.obolibrary.org/obo/CHEBI_>
6   PREFIX   chebilocal : <http://purl.obolibrary.org/obo/CHEBI_>
7   PREFIX   chebiremote : <http://purl.obolibrary.org/obo/CHEBI:>
8   PREFIX   taxon : <http://identifiers.org/taxonomy/>
9
10
11  SELECT DISTINCT ?acid ?acidLabel ?reaction ?reactionName
12  WHERE  {
13    ?acid rdfs :subClassOf * chebilocal :25754 .
14    OPTIONAL  { ?acid rdfs :label ?acidLabel . }
15
16    BIND (URI (REPLACE (str (?acid ), "_" , ":")) AS ?acidRemote ) .
17
18    SERVICE <https://www.ebi.ac.uk/rdf/services/sparql>  {
19      ?reactant  biopax3 :entityReference  ?acidRemote  .
20
21      ?reaction  biopax3 :left ?reactant  .
22      ?reaction  rdf :type biopax3 :BiochemicalReaction  .
23      OPTIONAL  { ?reaction  biopax3 :displayName  ?reactionName  . }
24
25      ?pathway  biopax3 :pathwayComponent  ?reaction .
26      ?pathway  biopax3 :organism  taxon :9606 .
27    }
28  }
```

**Figure 2.7.** *SPARQL query combining a local endpoint with ChEBI and a remote endpoint with Reactome and that enables all chemical reactions consuming an oxo-carboxylic acid or one of its direct or indirect subclasses in humans to be found*

COMMENTS ON FIGURE 2.7.– *This is a federated version of the query in Figure 2.6. The* SERVICE *clause specifies the URL of the remote endpoint (line 18) and the portion of the query to be sent thereto (lines 19–27). It should be noted that the variable* acidRemote *is common between the local and remote portion of the query. The ChEBI ontology and the Reactome database do not exactly employ the same prefixes for the URIs of the molecules, compelling them to be rewritten in an inelegant form (line 16).*

SPARQL makes it possible to create **federated queries**, including multiple endpoints, and to view their data as if they were a single virtual set of triples (Cheung et al. 2009; Kratochvíl et al. 2019). The SPARQL engine is then responsible for sending the query fragments to the relevant endpoints, and especially for integrating the results. This last step is potentially complicated because it may require a lot of unions and joins. Such queries have several advantages: they allow users to take full advantage of the complementarity of the databases (including for combining data from one of their projects with the knowledge bases) and relieve them of the complex integration phase. Several recent works have focused on the application of federated queries to life science data analysis (Djokic-Petrovic et al. 2017; Hasnain et al. 2017; Zaki and Tennakoon 2017; Lombardot et al. 2018). Figure 2.7 shows a federated version of the query represented in Figure 2.6 by combining an endpoint for Reactome and an endpoint for ChEBI.

There are several barriers that prevent the adoption of federated queries, and mainly performance issues (Abdelaziz et al. 2018). This is especially the case since the federated queries that would be required in the life sciences seem to be more complex than those typically considered in *benchmarks* (Saleem et al. 2019). This indicates in counterpoint that Semantic Web technologies can be used to address queries that were not previously considered. This is thus a highly challenging avenue for future research in the next few years, and advances in terms of data architecture and federated query engines will have a much broader impact than the life sciences sub-community of the Semantic Web.

## 2.4. Conclusion

This chapter has contributed to identifying the major role that data integration and analysis play in the life sciences, in order to make comprehensive analyses and move beyond ad hoc and partial analyses in each subfield. This has resulted in bringing forward requirements in terms of data representation, data structuring, queries and reasoning. These requirements are related both to data engineering and data science, and are resolutely computer based. It therefore seems appropriate to consider a general framework rather than ad hoc solutions. Semantic Web technologies have proven to be a viable solution to these requirements, and that their widespread adoption by the life sciences community is in progress. The success of this approach lies in particular in the fact that data analysis scenarios that were not previously possible can now be performed. The next challenges concern both the IT aspects through the improvement of query performance, especially federated queries, and the usability aspects to enable large-scale adoption by users who are experts in their field but not necessarily in the Semantic Web.

It has thus become clearly visible that, on the one hand, the life sciences are a privileged terrain for developing solutions, and, on the other hand, these solutions have a potential impact that also affects the entire Semantic Web community.

## 2.5. References

Abdelaziz, I., Mansour, E., Ouzzani, M., Aboulnaga, A., Kalnis, P. (2018). Lusail: A system for querying linked data at scale. *Proceedings of the VLDB Endowment*, 11(4), 485–498.

Al Kawam, A., Sen, A., Datta, A., Dickey, N. (2018). Understanding the bioinformatics challenges of integrating genomics into healthcare. *IEEE Journal of Biomedical and Health Informatics*, 22(5), 1672–1683.

Aldhous, P. (1993). Managing the genome data deluge. *Science (Washington)*, 262(5133), 502–503.

Anguita, A., García-Remesal, M., de la Iglesia, D., Maojo, V. (2013). NCBI2RDF: Enabling full RDF-based access to NCBI databases. *BioMed Research International*, 983805.

Aranguren, M.E.N., Bechhofer, S., Lord, P., Sattler, U., Stevens, R. (2007). Understanding and using the meaning of statements in a bio-ontology: Recasting the gene ontology in OWL. *BMC Bioinformatics*, 8, 57.

Bard, J.B.L. and Rhee, S.Y. (2004). Ontologies in biology: Design, applications and future challenges. *Nature Reviews Genetics*, 5(3), 213–222.

Baumgartner, W.A., Cohen, K.B., Fox, L.M., Acquaah-Mensah, G., Hunter, L. (2007). Manual curation is not sufficient for annotation of genomic databases. *Bioinformatics (Oxford, England)*, 23(13), i41–i48.

Bechhofer, S., Buchan, I., De Roure, D., Missier, P., Ainsworth, J., Bhagat, J., Couch, P., Cruickshank, D., Delderfield, M., Dunlop, I. et al. (2013). Why linked data is not enough for scientists. *Future Generation Computer Systems*, 29(2), 599–611.

Bellazzi, R. (2014). Big data and biomedical informatics: A challenging opportunity. *Yearbook of Medical Informatics*, 9(1), 8–13.

Bellazzi, R., Diomidous, M., Sarkar, I.N., Takabayashi, K., Ziegler, A., McCray, A.T. (2011). Data analysis and data mining: Current issues in biomedical informatics. *Methods of Information in Medicine*, 50(6), 536–544.

Belleau, F., Nolin, M.-A., Tourigny, N., Rigault, P., Morissette, J. (2008). Bio2RDF: Towards a mashup to build bioinformatics knowledge systems. *Journal of Biomedical Informatics*, 41(5), 706–716.

Berners-Lee, T., Hendler, J., Lassila, O. (2001). The Semantic Web. *Scientific American*, 284(5), 34–43.

Berners-Lee, T., Hall, W., Hendler, J.A., O'Hara, K., Shadbolt, N., Weitzner, D.J. (2007). A framework for web science. *Foundations and Trends in Web Science*, 1(1), 1–130.

Bizer, C., Heath, T., Berners Lee, T. (2009). Linked data – The story so far. *International Journal on Semantic Web and Information Systems*, 5(3), 1–22.

Blake, J.A. and Bult, C.J. (2006). Beyond the data deluge: Data integration and bio-ontologies. *Journal of Biomedical Informatics*, 39(3), 314–320.

Bodenreider, O. and Stevens, R. (2006). Bio-ontologies: Current trends and future directions. *Briefings in Bioinformatics*, 7(3), 256–274.

Brandizi, M., Singh, A., Rawlings, C., Hassani-Pak, K. (2018). Towards FAIRer biological knowledge networks using a hybrid linked data and graph database approach. *Journal of Integrative Bioinformatics*, 15(3), 20180023.

Bult, C.J. (2006). From information to understanding: The role of model organism databases in comparative and functional genomics. *Animal Genetics*, 37(Suppl. 1), 28–40.

Callahan, A., Cruz-Toledo, J., Dumontier, M. (2013). Ontology-based querying with Bio2RDF's linked open data. *Journal of Biomedical Semantics*, 4(Suppl. 1), S1.

Cannata, N., Merelli, E., Altman, R.B. (2005). Time to organize the bioinformatics resourceome. *PLoS Computational Biology*, 1(7), 0531–0533.

Cannata, N., Schröder, M., Marangoni, R., Romano, P. (2008). A Semantic Web for bioinformatics: Goals, tools, systems, applications. *BMC Bioinformatics*, 9(Suppl. 4), S1.

Chen, H., Yu, T., Chen, J.Y. (2012). Semantic Web meets integrative biology: A survey. *Briefings in Bioinformatics*, 14(1), 109–125.

Cheung, K.-H., Frost, H.R., Marshall, M.S., Prud'hommeaux, E., Samwald, M., Zhao, J., Paschke, A. (2009). A journey to Semantic Web query federation in the life sciences. *BMC Bioinformatics*, 10(Suppl. 10), S10.

Cimino, J.J. (1998). Desiderata for controlled medical vocabularies in the twenty-first century. *Methods of Information in Medicine*, 37(4–5), 394–403.

Cimino, J.J. (2005). In defense of the desiderata. *Journal of Biomedical Informatics*, 39(3), 299–306.

Cimino, J.J. and Zhu, X. (2006). The practical impact of ontologies on biomedical informatics. *Yearbook of Medical Informatics*, 124–135 [Online]. Available at: https://pubmed.ncbi.nlm.nih.gov/17051306/.

Clark, T., Martin, S., Liefeld, T. (2004). Globally distributed object identification for biological knowledge bases. *Briefings in Bioinformatics*, 5(1), 59–70.

Corby, O., Dieng-Kuntz, R., Faron-Zucker, C. (2004). Querying the Semantic Web with corese search engine. *Proceedings of the 16th European Conference on Artificial Intelligence (ECAI 2004)*, Valencia, August 22–27, 705–709.

Côté, R.G., Jones, P., Martens, L., Kerrien, S., Reisinger, F., Lin, Q., Leinonen, R., Apweiler, R., Hermjakob, H. (2007). The protein identifier cross-referencing (PICR) service: Reconciling protein identifiers across multiple source databases. *BMC Bioinformatics*, 8, 401.

van Dam, J.C., Koehorst, J.J., Schaap, P.J., Martins Dos Santos, V.A., Suarez-Diez, M. (2015). Rdf2graph a tool to recover, understand and validate the ontology of an rdf resource. *Journal of Biomedical Semantics*, 6, 39.

Djokic-Petrovic, M., Cvjetkovic, V., Yang, J., Zivanovic, M., Wild, D.J. (2017). PIBAS FedSPARQL: A web-based platform for integration and exploration of bioinformatics datasets. *Journal of Biomedical Semantics*, 8(1), 42.

Ferré, S. (2014). Expressive and scalable query-based faceted search over SPARQL endpoints. *The Semantic Web – ISWC 2014*, 8797, 438–453. doi: 10.1007/978-3-319-11915-1_28.

Goble, C. and Stevens, R. (2008). State of the nation in data integration for bioinformatics. *Journal of Biomedical Informatics*, 41(5), 687–693.

Goodman, A., Pepe, A., Blocker, A.W., Borgman, C.L., Cranmer, K., Crosas, M., Di Stefano, R., Gil, Y., Groth, P., Hedstrom, M. et al. (2014). Ten simple rules for the care and feeding of scientific data. *PLoS Computational Biology*, 10(4), e1003542.

Hasnain, A., Mehmood, Q., Sana E Zainab, S., Saleem, M., Warren, C., Zehra, D., Decker, S., Rebholz-Schuhmann, D. (2017). BioFed: Federated query processing over life sciences linked open data. *Journal of Biomedical Semantics*, 8(1), 13.

Hill, D.P., Adams, N., Bada, M., Batchelor, C., Berardini, T.Z., Dietze, H., Drabkin, H.J., Ennis, M., Foulger, R.E., Harris, M.A. et al. (2013). Dovetailing biology and chemistry: Integrating the gene ontology with the ChEBI chemical ontology. *BMC Genomics*, 14, 513.

Holford, M.E., McCusker, J.P., Cheung, K.-H., Krauthammer, M. (2012). A Semantic Web framework to integrate cancer omics data with biological knowledge. *BMC Bioinformatics*, 13(Suppl. 1), S10.

Juty, N., Le Novère, N., Laibe, C. (2012). Identifiers.org and miriam registry: Community resources to provide persistent identification. *Nucleic Acids Research*, 40(Database issue), D580–D586.

Kalderimis, A., Lyne, R., Butano, D., Contrino, S., Lyne, M., Heimbach, J., Hu, F., Smith, R., Stepán, R., Sullivan, J., Micklem, G. (2014). Intermine: Extensive web services for modern biology. *Nucleic Acids Research*, 42(Web server issue), W468–W472.

Kratochvíl, M., Vondrášek, J., Galgonek, J. (2019). Interoperable chemical structure search service. *Journal of Cheminformatics*, 11(1), 45.

Levesque, H.J. (2014). On our best behaviour. *Artificial Intelligence*, 212, 27–35 [Online]. Available at: https://www.sciencedirect.com/science/article/pii/S0004370214000356.

Livingston, K.M., Bada, M., Baumgartner, W.A., Hunter, L.E. (2015). Kabob: Ontology-based semantic integration of biomedical databases. *BMC Bioinformatics*, 16, 126.

Lombardot, T., Morgat, A., Axelsen, K.B., Aimo, L., Hyka-Nouspikel, N., Niknejad, A., Ignatchenko, A., Xenarios, I., Coudert, E., Redaschi, N., Bridge, A. (2018). Updates in rhea: SPARQLing biochemical reaction data. *Nucleic Acids Research*, 47(D1), D693–D700.

McCarthy, L., Vandervalk, B., Wilkinson, M. (2012). SPARQL assist language-neutral query composer. *BMC Bioinformatics*, 13(Suppl. 1), S2.

McMurry, J.A., Juty, N., Blomberg, N., Burdett, T., Conlin, T., Conte, N., Courtot, M., Deck, J., Dumontier, M., Fellows, D.K. et al. (2017). Identifiers for the 21st century: How to design, provision, and reuse persistent identifiers to maximize utility and impact of life science data. *PLoS Biology*, 15(6), e2001414.

Noy, N.F., Shah, N.H., Whetzel, P.L., Dai, B., Dorf, M., Griffith, N., Jonquet, C., Rubin, D.L., Storey, M.-A., Chute, C.G. et al. (2009). Bioportal: Ontologies and integrated data resources at the click of a mouse. *Nucleic Acids Research*, 37(Web server issue), W170–W173.

Pérez, J., Arenas, M., Gutierrez, C. (2009). Semantics and complexity of SPARQL. *ACM Trans. Database Syst.*, 34(3), 16:1–16:45.

Pierce, H.H., Dev, A., Statham, E., Bierer, B.E. (2019). Credit data generators for data reuse. *Nature*, 570(7759), 30–32.

Post, L.J.G., Roos, M., Marshall, M.S., van Driel, R., Breit, T.M. (2007). A Semantic Web approach applied to integrative bioinformatics experimentation: A biological use case with genomics data. *Bioinformatics*, 23(22), 3080–3087.

Rho, K., Kim, B., Jang, Y., Lee, S., Bae, T., Seo, J., Seo, C., Lee, J., Kang, H., Yu, U. et al. (2011). GARNET – Gene set analysis with exploration of annotation relations. *BMC Bioinformatics*, 12(Suppl. 1), S25.

Rigden, D.J. and Fernández, X.M. (2019). The 26th annual nucleic acids research database issue and molecular biology database collection. *Nucleic Acids Research*, 47(D1), D1–D7.

Rodríguez-Iglesias, A., Rodríguez-González, A., Irvine, A.G., Sesma, A., Urban, M., Hammond-Kosack, K.E., Wilkinson, M.D. (2016). Publishing fair data: An exemplar methodology utilizing phi-base. *Frontiers in Plant Science*, 7, 641.

Ruttenberg, A., Clark, T., Bug, W., Samwald, M., Bodenreider, O., Chen, H., Doherty, D., Forsberg, K., Gao, Y., Kashyap, V. et al. (2007). Advancing translational research with the Semantic Web. *BMC Bioinformatics*, 8(3) [Online]. Available at: https://doi.org/10.1186/1471-2105-8-S3-S2.

Sahoo, S.S., Bodenreider, O., Zeng, K., Sheth, A. (2007). An experiment in integrating large biomedical knowledge resources with RDF: Application to associating genotype and phenotype information. *Proceedings of the WWW2007 Workshop on Health Care and Life Sciences Data Integration for the Semantic Web*, Banff.

Saleem, M., Szárnyas, G., Conrads, F., Bukhari, S.A.C., Mehmood, Q., Ngonga Ngomo, A.-C. (2019). How representative is a SPARQL benchmark? An analysis of RDF triplestore benchmarks. *Proceedings of ACM Conference (WWW'19: The World Wide Web Conference)*, 1623–1633 [Online]. Available at: https://dl.acm.org/doi/10.1145/3308558.3313556.

Salvadores, M., Horridge, M., Alexander, P.R., Fergerson, R.W., Musen, M.A., Noy, N.F. (2012). Using SPARQL to query Bioportal ontologies and metadata. *Proceedings of the International Semantic Web Conference ISWC 2012*, 7650, 180–195 [Online]. Available at: http://iswc2012.semanticweb.org/.

Scharffe, F., Atemezing, G., Troncy, R., Gandon, F., Villata, S., Bucher, B., Hamdi, F., Bihanic, L., Képéklian, G., Cotton, F. et al. (2012). Enabling linked data publication with the datalift platform. *AAAI 2012, 26th Conference on Artificial Intelligence, W10: Semantic Cities*, July 22–26, Toronto.

Schulz, S., Balkanyi, L., Cornet, R., Bodenreider, O. (2013). From concept representations to ontologies: A paradigm shift in health informatics? *Healthcare Informatics Research*, 19(4), 235–242.

Shadbolt, N., Hall, W., Berners Lee, T. (2006). The Semantic Web revisited. *IEEE Intelligent Systems*, 21(3), 96–101.

Sima, A.C., Stockinger, K., de Farias, T.M., Gil, M. (2019). Semantic integration and enrichment of heterogeneous biological databases. *Methods in Molecular Biology*, 1910, 655–690.

Smedley, D., Haider, S., Durinck, S., Pandini, L., Provero, P., Allen, J., Arnaiz, O., Awedh, M.H., Baldock, R., Barbiera, G. et al. (2015). The BioMart community portal: An innovative alternative to large, centralized data repositories. *Nucleic Acids Research*, 43(W1), W589–W598.

Smith, B. (2005). New desiderata for biomedical terminologies. *Ontologies and Biomedical Informatics, Conference of the International Medical Informatics Association*.

Stephens, Z.D., Lee, S.Y., Faghri, F., Campbell, R.H., Zhai, C., Efron, M.J., Iyer, R., Schatz, M.C., Sinha, S., Robinson, G.E. (2015), Big data: Astronomical or genomical? *PLoS Biology*, 13(7), e1002195.

Stevens, R., Goble, C.A., Bechhofer, S. (2000). Ontology-based knowledge representation for bioinformatics. *Briefings in Bioinformatics*, 1(4), 398–416.

Stevens, R., Egaña Aranguren, M., Wolstencroft, K., Sattler, U., Drummond, N., Horridge, M., Rector, A. (2007). Using OWL to model biological knowledge. *International Journal of Human Computer Studies*, 65(7), 583–594.

Taboada, M., Martínez, D., Pilo, B., Jiménez-Escrig, A., Robinson, P.N., Sobrido, M.J. (2012). Querying phenotype-genotype relationships on patient datasets using Semantic Web technology: The example of cerebrotendinous xanthomatosis. *BMC Medical Informatics and Decision Making*, 12, 78.

Whetzel, P.L., Noy, N.F., Shah, N.H., Alexander, P.R., Nyulas, C., Tudorache, T., Musen, M.A. (2011). BioPortal: Enhanced functionality via new web services from the national center for biomedical ontology to access and use ontologies in software applications. *Nucleic Acids Research*, 39(Web server issue), W541–W545.

Wilkinson, M.D., Dumontier, M., Aalbersberg, I.J.J., Appleton, G., Axton, M., Baak, A., Blomberg, N., Boiten, J.-W., da Silva Santos, L.B., Bourne, P.E. et al. (2016). The fair guiding principles for scientific data management and stewardship. *Scientific Data*, 3, 160018.

Wimalaratne, S.M., Bolleman, J., Juty, N., Katayama, T., Dumontier, M., Redaschi, N., Le Novère, N., Hermjakob, H., Laibe, C. (2015). SPARQL-enabled identifier conversion with identifiers.org. *Bioinformatics*, 31(11), 1875–1877.

Yamaguchi, A., Kozaki, K., Lenz, K., Wu, H., Kobayashi, N. (2014). An intelligent SPARQL query builder for exploration of various life-science databases. *Proceedings of the 3rd International Conference on Intelligent Exploration of Semantic Data*, 1279, 83–94.

Zaki, N. and Tennakoon, C. (2017). BioCarian: Search engine for exploratory searches in heterogeneous biological databases. *BMC Bioinformatics*, 18(1), 435.

Zhang, Y., De, S., Garner, J.R., Smith, K., Wang, S.A., Becker, K.G. (2010). Systematic analysis, comparison, and integration of disease based human genetic association data and mouse genetic phenotypic information. *BMC Medical Genomics*, 3, 1.

# 3

# Workflows for Bioinformatics Data Integration

Sarah COHEN-BOULAKIA[1] and Frédéric LEMOINE[2]
[1] Université Paris-Saclay, CNRS, LISN, Orsay, France
[2] Institut Pasteur, Université Paris Cité, France

## 3.1. Introduction

Numerous bioinformatics experiments are performed daily in many laboratories around the world. Their results are stored in a multitude of databases, websites and publications, and follow different conventions, schemes and formats.

Providing uniform access to these highly heterogeneous, diverse and distributed data(bases) is the goal of **data integration**-based solutions that have been designed and implemented within the bioinformatics community for over 25 years. The goals and benefits of data integration from a biologist's perspective are numerous and include reduced costs (more focused maintenance), improved data quality (more reliable integrated data), new discoveries (combining complementary work) and faster discoveries made possible (by reusing instead of redoing).

---

For a color version of all figures in this chapter, see: http://www.iste.co.uk/froidevaux/biologicaldata.zip.

*Biological Data Integration*,
coordinated by Christine FROIDEVAUX, Marie-Laure MARTIN-MAGNIETTE and Guillem RIGAILL © ISTE Ltd 2023.

Today, the need for data integration in life sciences is still increasingly growing. However, experimentation throughput has dramatically changed over the last three decades – we have witnessed a shift from single experimental observations to gigabytes of sequences generated in a single day – and the scope of scientific questions being studied has widended – from individual genomes to populations, transcriptomes, proteomes, etc. – which now makes the problem of combining and managing biological data falls directly under the umbrella of Big Data management and more generally into the challenges of **data science**.

More importantly, it is the very perception of the problem of data integration in life sciences research that has changed: while early approaches focused on processing queries on heterogeneous and distributed databases, current approaches mainly emphasize instances (the data) rather than schemas (for instance, the tables that contain the data) and strive to reinstate human beings back in the loop. The data integration and data analysis phases are no longer juxtaposed but intertwined. The old model "first integrate, then analyze" is substituted by a new paradigm focused on the analysis and integration processes: "one integrates to better analyze, and one analyzes to better integrate".

The central object of this paradigm shift is the integration (and thus analysis) process itself: the bioinformatics data processing chain implemented in the form of a scientific workflow.

The objective of this chapter is to bring forward present difficulties and the existing solutions for the design and execution of bioinformatics processing chains. More specifically, section 3.2 presents the difficulties encountered and the needs that emerge in designing and executing a bioinformatics processing chain when implemented using scripts (in Python, for example). Section 3.3 introduces the concept of workflow, describes the principles behind workflow management systems, and introduces and compares several systems which are now widely used by the bioinformatics community. Section 3.4 describes a real use case of a bioinformatics processing chain and presents how it can be implemented using these different systems. Section 3.5 presents the unresolved (research) problems associated with this theme. Section 3.6 draws a conclusion.

## 3.2. Bioinformatics data processing chains: difficulties

In sciences that generate large amounts of data, such as bioinformatics, scientific results are typically produced using complex data processing chains which can take very large amounts of experimental data as input. These processing chains may consist of many steps, involve many tools and require considerable computational time.

For example, the 100,000 Genomes project[1] initiated by the UK government in 2012 has sequenced more than 75,000 patients suffering from rare diseases, as well as their families and patients suffering from cancer. The raw data reach 30 petabytes[2], and the data processing chains developed to analyze such a large volume of data are complex and greedy in terms of CPU hours.

Usually, these data processing chains are implemented using scripts (Bash, Perl, Python, etc.) that execute various tools, and thus constitute the link between the data, the processing, the tools and the execution environment. To this end, these scripts are used for (1) formatting, converting, cleaning and filtering the initial or intermediate data, (2) executing the tools in the appropriate order and (3) interacting with computing machines (cluster).

However, the development and use of such scripts lead to a number of difficulties, described in the following section, at different stages of the project lifecycle, namely (1) difficulties in the design and implementation of the data processing chain, (2) execution and reproducibility difficulties and (3) testing, maintenance, sharing and reuse-related issues.

## 3.2.1. *Designing a data processing chain*

The implementation issues related to the development of bioinformatics data processing chains in the form of ad hoc analysis scripts are of different kinds.

### 3.2.1.1. *Variety of languages and dependency on the software environment*

Many data processing chains resort to multiple languages and software environments. For example, sequencing data can be collected with a Bash script using the SRA Toolkit[3], the data format can then be converted with a Python script using the BioPython library (Cock et al. 2009) and a statistical analysis can be performed with an *R* script[4] requiring dedicated libraries.

These various languages and software environments must be unified by the developer, which makes the implementation of the processing chain more complex.

---

1. Available at: https-3.5pt://www.genomicsengland.co.uk/about-genomics-england/the-100000-genomes-project/.
2. Available at: https://fr.slideshare.net/dmontaner/100000-genomes-project.
3. Available at: https://www.ncbi.nlm.nih.gov/books/NBK242621/.
4. Available at: https://cran.r-project.org/.

### 3.2.1.2. *Testing and versioning*

Writing tests is an important part of good software development practices. Yet, for the development of analysis scripts, they prove to be difficult to design for different reasons: (1) it is necessary to isolate a dataset representative of the real data to test the analyses, (2) the lack of modularity of an analysis script can excessively complicate the testing of each step individually and (3) the lack of a dedicated testing framework makes the development of these tests particularly challenging.

Moreover, it is important to keep track of the different versions of the executed code, but above all to visualize these differences and to test the impact of changes made to a data processing chain. Development tools with these features are not currently capable of managing multilingual scripts.

## 3.2.2. Analysis execution and reproducibility

The execution of a processing chain and its reproducibility also raise a number of difficulties for its developer, who will have to look after the execution site, task scheduling, potential execution errors and the selective execution of certain steps into consideration.

### 3.2.2.1. *Dependency on the execution site*

First, analysts must comply with the ecosystem within which the analyses will be executed. They must design the data processing chain according to the site where the analysis will be executed: local machine, *cluster* or cloud, for example. The dependency on the environment of the developed scripts is therefore severe, and their genericity and portability is not guaranteed. Considering this context, development complexity is greatly increased.

### 3.2.2.2. *Scheduling*

The implementation of a data processing chain in the form of a script is generally linear. The steps follow one another, and it is therefore difficult to address the execution in a parallel way according to the data flow. Two levels of scheduling and parallelization are possible and are the responsibility of the developer of the data processing chain:

– the scheduling and parallelization of the execution of the different steps between them: if step 2 does not depend on the result of step 1, then the two steps can be executed in parallel if the resources allow it (in terms of CPUs and RAM);

– the scheduling and parallelization of the execution of each step: if a step must be executed several times with different input data, then these executions can be parallelized.

These two difficulties are becoming unavoidable with the amount of data processed today and are, if necessary, partly under the responsibility of the scheduler of the computational cluster, such as SLURM (Jette et al. 2002). Nonetheless, they considerably complicate the development of analysis scripts.

### 3.2.2.3. Error management

The management of execution errors is one of the difficulties related to the development of analysis scripts. As a matter of fact, during the execution of a processing chain, many errors can occur at different levels, and in particular errors related to (1) the implementation (bug in the data processing chain, bad parameterization of the tools) and (2) the environment (*cluster* failure), temporary inaccessibility to data or tools, bad scaling of the computing machines: RAM, CPUs, storage.

A processing chain must then be able to propose different responses to these errors. First of all, in the case of an implementation bug, the processing chain should be stopped. Then, it is possible to ignore the error if it is expected (empty data on input leading to empty data on output for example). In the case of a computational node failure or the temporary inaccessibility to data for example, it is desirable to be able to run again the failed task without incidence on the other tasks, executed or in progress. Finally, when poorly assessing what resources should be allocated to a task (not enough CPUs or RAM), it is important to be able to rerun this task asking for more resources, for example.

In all these configurations, the management of error cases constitutes a new difficulty for the development of the analysis script.

### 3.2.2.4. Selective rerun

In an exploratory project, analyses are often incrementally developed, in other words, by adding or changing analysis steps over time. When modifying a processing chain, if it is rerun with the same input data, it is desirable to keep the already computed results of the entire sequence of unmodified steps that are not dependent on the modified steps. The only steps that will thus be rerun are the new, modified steps which generate errors and their downstream steps.

This selective execution is all the more important as the steps in question are costly in terms of computing time. On the other hand, it is difficult to achieve with a simple analysis script, and in this case it is generally accomplished manually by temporarily deleting from the script the calls to the tasks that have already been executed.

### 3.2.3. *Maintenance, sharing and reuse*

Once the data processing chain has been designed, executed multiple times, it is considered to be robust and can, for example, be employed routinely on an analysis platform. This eventually leads to the need for maintenance and for sharing the data processing chain, which implies its reuse.

#### 3.2.3.1. *Maintenance*

The maintenance of processing chains developed in the form of analysis scripts is sensitive due to the cross-language aspects of the analysis, its lack of modularity and its strong dependence on the execution site.

Technologies and hardware evolve, and a script that runs successfully at a given time may become obsolete some time later, when the hardware environment has changed (e.g. a particular hardware environment might no longer be available, or its scheduler has changed), or the software environment has changed (e.g. the tools or the same versions are no longer available, etc.).

It is therefore necessary to be able to maintain the processing chain without too much difficulty in order to overcome these different points.

#### 3.2.3.2. *Sharing and reuse of processing chains*

Sharing processing chains implies the possibility of reusing them over time, possibly by a third party. The problems that arise are of the same order as for maintenance, since the challenge is to make sure that the chain can be rerun on different hardware and software environment, which requires installing the correct versions of the tools and all the dependencies, adapting the chain to another execution environment (e.g. a computer running under another system), and proper access to the data.

It is therefore important here to keep track of the versions of the multiple tools executed, their parameters and their environment. This task is difficult and tedious in the case of a processing chain implemented in a linear script. This tendency is further exacerbated by the increase in time of the number of processing chains and the number of tools and their versions that must then be managed and stored.

Finally, in a cumulative research framework, it might be desirable to be able to reuse only a subpart of pre-existing data processing chains. This partial reuse is only possible if (1) the analysis steps are sufficiently independent of each other (modularity), (2) the steps are sufficiently documented and (3) the code, environment and data are accessible. The lack of genericity and modularity of a data processing chain implemented in a script does not allow for easy extraction and execution of specific steps, unless a significant development effort is made (Sandve et al. 2013).

## 3.3. Solutions provided by scientific workflow systems

Considerable effort has been dedicated over the past two decades to provide bioinformaticians with approaches to guide them toward better automation of their data analysis processing chains. In particular, **scientific workflow systems** have been designed with the promise of assisting bioinformaticians at multiple levels, especially in designing, developing, monitoring, executing, keeping track of data used and produced, reproducing, and sharing their data analyses (Pradal et al. 2017).

These systems capture the exact methodology of a data processing chain in a **workflow**. The **specification** of a workflow is defined by the sequence of bioinformatics tools to be executed (namely, the tools and the order in which they may be executed). **The execution** of a workflow is defined by a set of input data, the definition of tool parameters and the set of data produced (and consumed in turn) by the various tools in the workflow. A workflow specification and its execution are generally represented by graphs where the nodes are the tools and the links represent the data flow. We will see in section 3.3.1, through concrete examples, that according to the workflow systems the representations can vary and that some systems only allow the specification to be visualized, whereas others only allow the execution of the workflow to be visualized. The **runtime environment** of a workflow captures the workflow dependencies on its hardware and software environment.

A wide variety of systems such as Galaxy (Goecks et al. 2010), Snakemake (Köster and Rahmann 2012) and Nextflow (Di Tommaso et al. 2017) have reached a level of maturity allowing them to be used daily by scientists to manage their bioinformatics data analysis. Taverna (Wolstencroft et al. 2013) is the pioneering system of the domain, no longer maintained.

Section 3.3 introduces the principles associated with workflow systems and presents an overview of systems and tools widely used in bioinformatics.

### 3.3.1. *Fundamentals of workflow systems*

Two types of bioinformaticians are targeted by workflow systems: **user-developers** of new analysis methods whose role is to implement new analysis methods and make them available to the community (*power-users*) and **end-users** of systems (*end-users*) who design their analyses by reusing and chaining "bricks" that already exist (Cohen-Boulakia and Leser 2011). These two types of users of workflow systems are brought to work together. In large bioinformatics institutes or centers, dedicated teams have been recruited to manage analysis platforms equipped with workflow systems. Some systems (such as Snakemake) are oriented toward user-developers only, who have then the additional responsibility of presenting and discussing the results obtained by workflow executions with end-users.

In the rest of this section, we describe the main tasks of these users by distinguishing two phases: workflow design and execution.

#### 3.3.1.1. Workflow design: encapsulation and modularization

The role of the users-developers is to make the new analysis methods robust and therefore reusable.

More specifically, they are working on the modularization and the encapsulation of analysis methods into the workflow system.

##### 3.3.1.1.1. Encapsulation

Encapsulation consists in representing each step of the workflow in a unified implemented form, dedicated to the workflow system. Each of these steps can contain a script (in Python or Java, for example) describing an analysis or the call to a bioinformatics tool (installed on a machine, web service, etc.). The objective is to define an interface (in the implementation sense) as simplified as possible: inputs (data types and parameters) and outputs in a language and format recognized by the workflow system.

In this domain, a number of initiatives have been launched to try to homogenize the languages used by workflow systems in order to encapsulate them. IWIR was one of the first workflow specification languages (Plankensteiner et al. 2011) extended with a conceptual language (Cerezo and Montagnat 2011) for annotating scientific tasks. Since 2016, a *consortium* of multiple organizations and workflow system vendors have introduced CWL[5] (for Common Workflow Language) (Amstutz et al. 2016), with the purpose of promoting the portability of workflow specifications. CWL is currently under development. More than a simple workflow representation language, the challenge is to allow the execution of analyses implemented in different systems (workflow interoperability).

##### 3.3.1.1.2. Modularization

Modularization consists in turning these analysis steps into **modules**, independent from the other steps of the workflow and therefore reusable. These modules can be documented at the level of their inputs/outputs and their functionalities, in a more or less formal way: free text or annotation using terms from dedicated ontologies. These modules can then be reused and chained together by end-users to form workflows.

---

5. Available at: https://github.com/common-workflow-language.

Some workflow systems include graphical interfaces where users can select modules from libraries and import them into a work environment to assemble them. In turn, workflows can be grouped in workflow repositories and be annotated and documented to be made available to the community and then reused. Another important aspect related to modularization and encapsulation is the possibility in most systems to design sub-workflows, defined as workflows whose complexity is hidden by encapsulating it in unitary and reusable modules.

A number of warehouses allow scientists to share and publish their workflows for the benefit of the scientific community. These warehouses are characterized by the workflow systems of their workflows and the functionalities they offer to their end-users (annotations using ontologies versus simple text, creation of user groups with preferred workflows, etc.). We can mention CrowdLabs (Mates et al. 2011), SHIWA (Korkhov et al. 2012), the warehouses proposed with Kepler and Galaxy (Blankenberg et al. 2014) and probably the most visible and the most advanced in terms of functionalities: myExperiment (Roure et al. 2009).

In addition to workflows warehouses, there are warehouses, or catalogs of software or tools to share information on the bioinformatics tools underlying the steps of the workflows. Biocatalogue (Bhagat et al. 2010) is the pioneering catalog of the domain and is now included in Bio.tools (Ison et al. 2016), a larger reference catalog.

Both types of warehouses (of workflows or tools) are accompanied by annotations, free text or controlled vocabularies, describing the workflows or the tools. It should be noted that Bio.tools uses the concepts of ontologies of the domain, such as EDAM[6] (Ison 2013) to annotate the functionality of a given tool as well as its inputs, outputs and parameters.

### 3.3.1.2. *Workflow execution: transparency, optimization and traceability*

From the end-users' point of view, the execution of a workflow should be as easy as possible: once the input data and the tool parameters have been specified, the workflow should be able to execute the different modules that generate intermediate and final data. Rerunning the same workflow is an important feature that should be made easier for end-users. In addition, current workflow systems also offer the possibility to rerun parts of the workflow (only the parts affected by a change).

As a result, user-developers are responsible for ensuring three fundamental tasks: transparency of workflow execution, workflow execution optimization and traceability of the data generated and produced during the execution.

---

6. Available at: http://edamontology.org/.

### 3.3.1.2.1. Transparency of workflow execution

In order to ensure the transparency of workflow executions, in addition to module encapsulation, it is necessary to capture the execution environment, that is, to keep track of the dependencies of the workflow modules with the environment required for its execution. These dependencies can be of various kinds: libraries installed on the executing machine, size of the memory required, proper versions of the tools, etc.

Several approaches have been proposed to preserve the workflow execution environment and its scientific context. Virtualization technologies, such as VMware[7], KVM[8], VirtualBox[9] and Vagrant[10], can be used to package or even "freeze" the workflow execution environments. These techniques are expensive because they require the entire runtime environment to be copied (including the operating system).

Solutions based on **containers** represent remarkable, lightweight alternatives. They capture only the specific dependencies required by applications and share the low-level components provided by the operating system. The containers are accompanied by their build **recipes**, in the form of Docker files that allow their composition, management and sharing. We will mention OpenVZ[11], LXC[12], or more massively used in bioinformatics Docker (Boettiger 2015), Singularity (Kurtzer et al. 2017) and Conda[13].

Other approaches such as ReproZip (Chirigati et al. 2013) and CDE (Guo 2012) make it possible to capture the command line history and associated input and output files, and "condition" them as workflows.

### 3.3.1.2.2. Optimizing workflow runtimes

Regarding optimization, the challenge is to deal with data analyses which are increasingly costly in terms of time and space, and involve considerable amounts of data. Therefore, a large majority of workflow systems offer users a simplified interface to distributed environments such as *clusters* or clouds. User-developers are in charge of parameterizing workflow systems to ensure the link between the workflow and the execution machine(s).

---

7. Available at: https://www.vmware.com/.
8. Available at: https://www.linux-kvm.org.
9. Available at: https://www.virtualbox.org/.
10. Available at: https://www.vagrantup.com/.
11. Available at: https://openvz.org/.
12. Available at: https://linuxcontainers.org/.
13. Available at: http://conda.pydata.org.

Several execution levels are then involved: the triggering of the execution of a task/module by the workflow system; the triggering of the execution of this task by the execution environment (a computing cluster, for example); and the practical execution of the task by the final machine. Each of these three levels is associated with a scheduler. Workflow systems are equipped with schedulers whose role is to organize the execution of tasks (the choice of the order and the right level of parallelization of certain modules or of the entire workflow); the execution environment (the cluster) executes tasks when resources are available and maximizes their parallelization; and the operating system of the execution machine chooses the order in which the processes are potentially executed in parallel.

Optimization of workflow execution is all the more important when the execution environment is dematerialized. The extreme case being the execution of tasks in the cloud, on machines that start on demand, according to the workflow needs.

Workflow systems with an emphasis on distributed scheduling have been studied in detail in Juve et al. (2013).

### 3.3.1.2.3. Workflow execution traceability

An increasing number of workflow systems are equipped with provenance modules, which automatically keep track (in databases) of the data used and generated during the analysis, and thus provide good traceability of the execution. This information is crucial to understand the results (biological or bioinformatic) obtained by the analysis so that it can be reproduced. This information can also be used to re-execute only part of a workflow (only the modified part). The main workflow systems use execution traces to resume the execution where it stopped if an execution error occurred, or if a step was added or modified. For this, systems can rely, for example, on a unique identifier associated with the execution of each step[14]. This identifier can be generated by hashing methods from the inputs, the analysis scripts and their parameters, the outputs and the execution environment. During a new run, the steps whose identifier has changed (as well as downstream steps) are then rerun.

Recent efforts have been deployed by major scientific workflow system vendors to adopt an open provenance model, resulting in the W3C PROV recommendation. PROV provides a very generic model for exchanging provenance data on the Web. PROV does not explicitly provide all the concepts needed to model workflow executions. As a result, several PROV-compliant languages coexist: OPMW[15],

---

14. Available at: https://www.nextflow.io/blog/2019/demystifying-nextflow-resume.html.
15. Available at: http://www.opmw.org/model/OPMW/.

wfprov[16], provwf (Costa et al. 2013) and ProvONE[17]. These languages differ in the type of information they capture but agree on the PROV concepts used to model both the basic concepts and relationships of workflow executions.

### 3.3.2. Workflow systems

We have chosen to summarize four systems: Taverna (Wolstencroft et al. 2013), Galaxy (Giardine et al. 2005; Goecks et al. 2010), Snakemake (Köster and Rahmann 2012) and Nextflow (Di Tommaso et al. 2017) as the main representative of the diversity of scientific workflows used for analyzing bioinformatics data. We limit the systems presented here to one representative per type of system. Systems such as Kepler (Ludäscher et al. 2006), Wings (Gil et al. 2010) ou Knime (Berthold et al. 2009) share common functionalities with one or more of the workflow systems described in this section.

A summary and comparative information about the systems used for workflow specification, execution and environment management are provided in the following tables, respectively: Tables 3.1, 3.2 and 3.3. In these tables, "/" means that the criteria in question are not supported by the system.

#### 3.3.2.1. Taverna

Taverna, first released in 2013, is the scientific workflow system that made it possible for the first time composing and chaining bioinformatics tools in a user-friendly way.

#### 3.3.2.1.1. General overview

For workflow specification, Taverna has its own workflow description language called SCUFL2 (Soiland-Reyes et al. 2014), an XML format that extends W3C PROV. The modules that make up a Taverna workflow encapsulate local programs (*script bean*, *R-script*) or remote programs (REST or cloud). A user can specify a workflow or reuse existing workflows as sub-workflows. Taverna integrates search functionalities allowing workflows and tools from myExperiment and Biocatalogue to be found.

Regarding workflow execution, Taverna is designed to capture the provenance of workflow executions using Taverna-PROV, a model that extends PROV and WfPROV[18]. Taverna provides a summary of the results of the execution: inputs and outputs of modules, data artifacts used and generated.

---

16. Available at: http://purl.org/wf4ever/wfprov.
17. Available at: http://vcvcomputing.com/provone/provone.html.
18. Available at: http://purl.org/wf4ever/wfdesc*5C*.

|  | Taverna | Galaxy | Nextflow | Snakemake |
|---|---|---|---|---|
| Graphical interface | Yes | Yes | No | No |
| Workflow annotations | Free text (*tags*) describing the workflow and its steps | Free text (*tags*) describing the workflow and its steps | / | / |
| Sub-workflows | Yes | Yes | Yes | Yes |
| Specification language | Taverna XML (SCUFL2) ontology | JSON dedicated format | Dedicated format | Dedicated format |
| Interoperability | Extends PROV | CWL compatible | / | CWL compatible |
| Workflow sharing | Yes | Yes | Yes | Yes |
| Workflow versioning | Yes | No | Yes (git) | Yes (git) |
| Tools | | | | |
| Scripts | Java, R | / | Multilanguage | Multilanguage |
| Command-line | Yes | Yes | Yes | Yes |
| Web services | Yes | No | No | No |

**Table 3.1.** *Features related to the workflow specification*

|  | Taverna | Galaxy | Nextflow | Snakemake |
|---|---|---|---|---|
| Cluster/Cloud | / | SGE, LSF, Slurm, Torque, PBS, HTCondor, etc. | SGE, LSF, SLURM, Torque, PBS, HTCondor, Ignite, Kubernetes, AWS Batch, Google Pipelines, etc. | DRMMA, cluster engines that support shell scripts and access to common file systems |
| Scheduling | Yes | Yes | Yes | Yes |
| Error handling levels | / | Stop | Ignore, Retry, Stop | Stop, Retry |
| Selective resume | Yes | Manual | Yes | Yes |
| Execution traces | TavernaPROV | Binary & JSON | Text dedicated format | Dedicated format |
| Standard used | PROV | / | / | / |
| Presentation | GUI for displaying and exploring the executions | GUI for exploring, exchanging and reusing histories | Reports on execution traces, performances and execution graphs | Reports on execution traces, performances and execution graphs |

**Table 3.2.** *Features related to the workflow runtime*

At the execution environment level, efforts are underway to provide Taverna with links to container-based virtualization tools such as Docker. Taverna includes a plug-in allowing it to run in a computational grid environment. With respect to capturing the broader scientific environment, Taverna workflows can be integrated into reproducible research objects such as *Research Objects* (Belhajjame et al. 2015).

|  | Taverna | Galaxy | Nextflow | Snakemake |
|---|---|---|---|---|
| **System environment capture** | Docker (very soon) | Conda, Docker and Toolshed | Docker, Singularity, Modules and Conda | Docker, Singularity and Conda |
| **Scientific environment capture** | Research Objects | Research Objects and ISA | / | Archive |

**Table 3.3.** *Software and hardware environment*

### 3.3.2.1.2. Limitations

Taverna has undoubtedly been the most innovative workflow system. Today, it is less used than the other three systems described in section 3.3. In particular, its architecture is very much tied to the use of web services, and this dependency on **volatile** third-party services causes problems at execution if the services are not available. Taverna workflows also tend to become complex by containing a large number of **adaptive modules** used to perform data format transformations. Taverna also has limited support for virtualization.

### 3.3.2.2. *Galaxy*

Galaxy is a workflow system dedicated to bioinformatics. It is very popular in the bioinformatics community due to its user-friendly interface that does not require any development knowledge. The first version of the system was released in 2009.

### 3.3.2.2.1. General overview

Galaxy workflows are composed of modules (called *tools*) described in files (*wrappers*) in XML format by the developers. These files contain the software called inputs, outputs and parameters. A workflow consists of modules chained together (by their input–output relationships) and parameterized by the workflow developer. Galaxy workflows are specified in a dedicated JSON format, compatible with the CWL language (*Common Workflow Language*). Tools are gathered into *toolsheds*, constituting a repository (Blankenberg et al. 2014) capable of managing tool versions as well as their dependencies for automated local deployment. Galaxy also provides *Galaxy Pages* for documenting and sharing workflows and tools.

Galaxy keeps track of its workflow executions using *histories*, a user-specific workspace containing tool configuration settings, module inputs and outputs (Blankenberg et al. 2014) and final results produced. A graphical interface for exploring and sharing workflow executions is provided by Galaxy.

Galaxy controls the context and execution environment by using a number of external solutions in order to resolve tool dependencies, such as Docker and Conda. Concerning the scientific packaging, Galaxy workflows can be packaged into reproducible research objects and ISA experiments (Gonzalez-Beltran et al. 2014).

### 3.3.2.2.2. Limitations

Galaxy currently has two main limitations. The first is related to the technical complexity of the system, which relies on specific software and reference datasets which must be locally available with the appropriate version to allow the workflow to be repeated. While the best practices in terms of system management make it possible to avoid such problems, using Toolsheds can result in redundant maintenance tasks, when the same tools must be available outside the Galaxy environment. Secondly, Galaxy has some limitations with respect to provenance. *Histories* in Galaxy record provenance information, but they are not based on a reference schema or vocabulary. The ability to query and export provenance information is limited.

### 3.3.2.3. *Snakemake*

Snakemake is a workflow system widely used in bioinformatics, rather intended for bioinformaticians capable of programming. The first version of the system was released in 2012.

### 3.3.2.3.1. General overview

Snakemake is implemented in Python. Unlike the workflow systems presented in sections 3.3.2.1 and 3.3.2.2, it does not include a graphical interface. Snakemake workflows are similar to GNU Make[19] build files. They consist of a set of rules rather than modules. Each rule specifies (1) one or more targets, the files that will be created when the rule is triggered; (2) the rule dependencies, the files that are necessary for it to be triggered; and (3) the commands (or scripts) that will be executed when it is triggered. Finally, a workflow specifies a global target (a result file expected in output, for example), and the rules directly and indirectly necessary to its production will then be triggered recursively. A Snakemake workflow is a graph whose nodes are the rules linked together by the dependencies between their left part (targets) and their right part (dependencies). Snakemake workflows are specified in a language that is not based on a standard but on an internal language.

For execution, just like GNU Make, Snakemake loads the workflow into memory and deduces the execution order of the rules based on the final target and the dependency relations between the rules to produce this target. Once the execution

---

19. Available at: https://www.gnu.org/software/make/.

graph is built, Snakemake can then execute the rules in the right order, parallelizing the steps when possible.

In terms of execution environment each rule can be executed locally or on a *cluster*, in a Docker, or a Singularity container, or a Conda environment that will be downloaded on demand.

#### 3.3.2.3.2. Limitations

Snakemake is based on rules that take file names as targets and dependencies. It is therefore the developer's responsibility to name the temporary files and output files in a way that meets the requirements of the workflow rules. In addition to naming the files, the developer has to consider the tree structure of the input and output files, which must not interfere between several executions of a rule.

Secondly, there is no tool repository (as Galaxy offers). Being compatible with Docker, Singularity and Conda, the developer then has to ensure that the workflow perfectly controls the execution environment for each rule.

Finally, Snakemake does not provide provenance traces represented in today's recognized standards (such as PROV), but in its own format.

### 3.3.2.4. *Nextflow*

Nextflow is one of the most recent workflow-based systems, implemented in Groovy (Koenig et al. 2007). Just like Snakemake, its use is targeted at bioinformaticians programmers. Its first version dates from 2013.

#### 3.3.2.4.1. General overview

Nextflow does not present a graphical interface, and Nextflow workflows are specified using a dedicated language in a text file. A Nextflow workflow is composed of two types of steps: modules (*process*) that encapsulate tools (Shell, Groovy, Python, R, etc.) and operators (*operators*) that implement basic data manipulation functions, internal to Nextflow. Nextflow modules are responsible for performing scientific computations and are defined by their inputs and outputs and the calculation to be performed.

Workflows are executed in a terminal in the order defined by the input–output relationships of the modules and operators. If a module is added or changed, or if the input data are modified, it is possible to specify that only the calculations affected by the change be rerun. Once the workflow is executed, the provenance information consists of tags associated with each module in the output log files.

Nextflow supports tool dependency management via Docker, Environment Modules (Furlani 1991), Singularity and Conda. Each module can specify in which container it should be executed or which environment it should load before execution. Nextflow will then ensure that the containers or environments needed to run the workflow are automatically downloaded. Modules can also specify how the computation will be executed: locally, on a cluster, or in a cloud.

It is possible to automatically download and execute workflows hosted on services such as GitLab, GitHub, BitBucket and so on.

### 3.3.2.4.2. Limitations

Nexflow has two main limitations. Firstly, since there is no central repository of tools, the developer has to implement the workflow by specifying as precisely as possible the software and hardware dependencies of the workflow with its environment. Secondly, the information related to execution and specification is not expressed in a standardized way (such as PROV) and therefore it is not possible to explore and exploit the provenance information.

## 3.4. Use case: RNA-seq data analysis

Workflow systems are increasingly used in the bioinformatics field, especially for the analysis of sequencing data. In this field, workflows consist of common steps (data cleaning, *mapping*, etc.) and steps specific to each study (statistical analysis, for example).

In this section, we introduce an RNA-seq data analysis that we have reimplemented and run with three major bioinformatics workflow systems.

### 3.4.1. *Study description*

#### 3.4.1.1. *Bioinformatics data analysis*

The use case considered here concerns the analysis of data published by Harbour et al. (2013) and reused by Furney et al. (2013). This analysis is representative of RNA-seq bioinformatics analyses, it uses a massive available dataset and the steps of the analysis are well described.

The objective of this study is to evaluate the impact of a mutation in the SF3B1 gene on gene expression in a group of patients suffering from uveal melanoma. The SF3B1 gene is a splicing factor and therefore plays an important role in the modification of RNAs (transcribed from genes) leading to the production of messenger RNAs that can then be translated into proteins (see Figure 3.1).

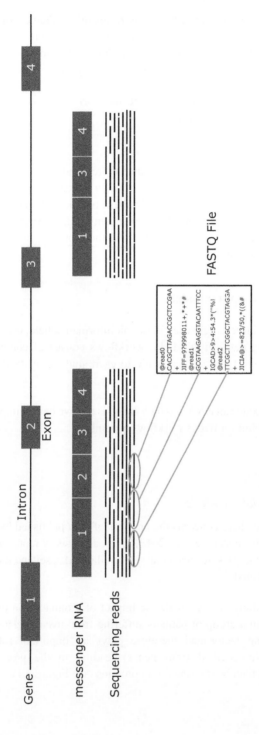

**Figure 3.1.** *Representation of alternative splicing and RNA-seq. Splicing makes it possible to produce a mature messenger RNA (without introns) from a pre-messenger RNA (with introns). Through this mechanism, a gene can generate several different messenger RNAs (alternative splicing), whose proteins will potentially have different functions*

The objective here is to study the impact of a mutation in the SF3B1 gene on RNA splicing in patients with uveal melanoma, one group of whom possesses the mutated version of the gene and another group who does not.

In this study, we have access to an RNA-seq dataset making it possible to qualitatively (presence of certain messenger RNAs) and quantitatively (level of presence) measure the RNA content.

For each sample, we have files containing sequence fragments, or reads (e.g. CATGCTCCCCCGTAT), and their quality (confidence in base identification). Since each read does not cover the whole messenger RNA from which it originates, this read is mapped onto the human reference genome (GRCh37/hg19[20]) to find the gene and the messenger RNA from which it originates.

### 3.4.1.2. *Implementation of the data processing chain*

The data processing chain conducting the data analysis is illustrated in Figure 3.2. The first step consists of downloading the data from *Sequence Reads Archive*[21] (SRA), a public sequencing data repository hosted by the *National Center for Biotechnology Information*[22] (NCBI) using the SRA toolkit.

The reads of each sample will then be aligned to the reference genome (GRCh37/hg19) with the STAR tool (Dobin et al. 2013), the outputs will be indexed with Samtools (Li et al. 2009) and the DEXSeq tool (Anders et al. 2012) will then count the number of reads mapping on each exon, which constitutes an approximation of the number of messenger RNAs possessing each exon.

The counts of each exon from all samples are then analyzed with DEXSeq in order to list the exons that are preferentially included or excluded from the mRNAs between the two groups of patients.

These six steps are a good sample of the diversity of languages that can be encountered when developing this kind of data processing chain. The download of the reference human genome is carried out with a Bash command; SRA Toolkit, STAR and Samtools are specific tools called from the command line; counting with DEXSeq is done with a python script, and DEXSeq statistical analysis is achieved with an R script.

---

20. Available at: http://hgdownload.soe.ucsc.edu/goldenPath/hg19/bigZips/.
21. Available at: https://www.ncbi.nlm.nih.gov/sra.
22. Available at: https://www.ncbi.nlm.nih.gov/.

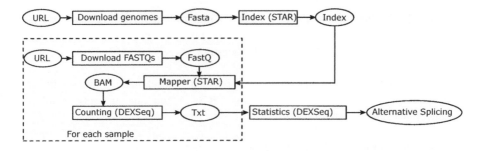

**Figure 3.2.** *Schematic representation of the data processing chain that analyzes RNA-seq data. Rectangles represent analysis tasks to be performed while ovals represent data types. A subset of the tasks must be performed multiple times, represented by a dotted rectangle*

### 3.4.2. From data processing chain to workflows

The data processing chain described above was implemented using the three most widely used bioinformatics workflow systems: Nextflow, Snakemake and Galaxy. These three implementations are available on GitHub[23].

The implementation of the data processing chain into a Nextflow workflow is presented in Figure 3.3 (its specification). The representation of the specification in Nextflow consists of nodes which are either analysis steps (in the form of ovals) or Nextflow specific operators (small circles), linked by arcs indexed by the name of the files that are produced. This workflow consists of 10 steps, mixing calls to Bash commands (data download), tool executions (sequence alignment to the genome, sequence counting on the exons) and R scripts (statistical analysis). Nextflow makes it possible to mix many languages within a single workflow without difficulty, and to very easily interface the execution on the computing cluster. Similarly, the Nextflow workflow resorts to Singularity containers at each step, which enables the control of the execution environment as well as the versions of the tools.

The implementation of the processing chain as a Snakemake workflow consists of 11 rules, encapsulating each of the major steps of the analysis. The additional rule compared to the Nextflow workflow defines the general target of the workflow (the final result the workflow should generate). In addition to the visualization of the specification, Snakemake provides a visualization of the execution graph, a fragment of which is shown in Figure 3.4. In this representation, the nodes represent the execution of the rules (nodes of the same color symbolize different executions of the

---

23. Available at: https://github.com/fredericlemoine/workflows_bioinfo_iste.

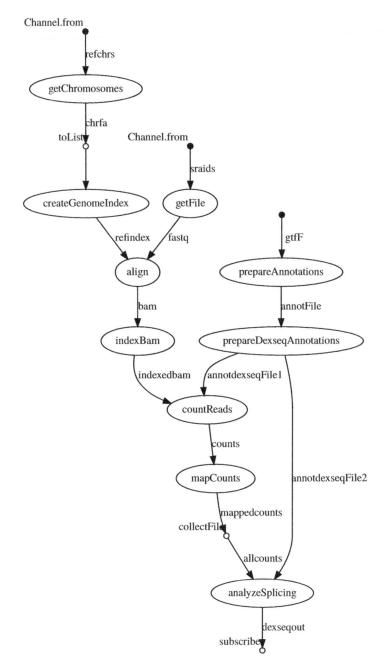

**Figure 3.3.** *Representation of the workflow specification by Nextflow*

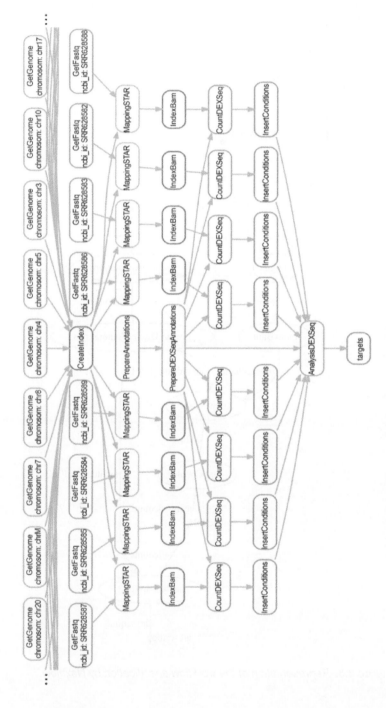

**Figure 3.4.** *Representation of the execution of the workflow by Snakemake*

same rule), and the arcs represent the dependency relations between the inputs and outputs of the rules. The workflow was executed on a cloud (8-core machines, 32G RAM, pre-installed with Bioconda, Docker, etc.) whose configuration was simplified by interfacing with Snakemake. Figure 3.4 also highlights the steps that can be executed in parallel by Snakemake, and the tasks that must wait for the previous tasks to finish executing. This scheduling is managed by Snakemake.

Finally, the data processing chain was also implemented as a Galaxy workflow, consisting of 10 modules. Figure 3.5 represents a screenshot of the workflow in the Galaxy interface. For Galaxy, the genome was made available to the system locally, therefore the download of the reference genome is not represented in the workflow. A Galaxy server can also provide access to a set of pre-downloaded genomes, such that users do not have to do this. The implementation of this workflow is different from that of Snakemake and Nextflow, as it must accommodate the modules already present on the Galaxy server being used. For example, the MapCount module in the Nextflow workflow, which adds the condition to each sample (mutated/unmutated), is no longer needed in the Galaxy workflow. As a matter of fact, the DEXSeq module can be used to provide the list of samples of each group. Similarly, some tools make available multiple outputs and only a sub-part of these outputs are connected to the rest of the workflow steps, because they are not needed for our purpose. The Galaxy workflow was run in a Galaxy instance itself running in a Docker container in which the tools were installed from the Toolshed, allowing to control the environment as well as tool versioning.

### 3.4.3. *Data processing chains implemented as workflows: conclusion*

The data processing chain, which includes different types of steps and makes use of various languages (R, Python, tools, etc.), has been implemented in three workflow systems that are very different both technically and in terms of usage. These implementations are configurable in multiple execution sites (local, *cluster*, etc.) and are able to exploit system schedulers (e.g. Snakemake). More precisely, for Snakemake and Nextflow, a configuration file (an example of which is given in Figure 3.6) placed next to the workflow can specify the type of machine that will execute it, the needs of the different steps of the workflow in terms of CPU and memory, and the Docker or Singularity containers in which each of the steps will run. In Galaxy, these configurations are managed by the Galaxy server. The partial rerun functionality could be exploited and was particularly useful in a context where the computation of some steps represents several hours. The implementation with low dependence on the execution site and "aware" of software dependencies simplified error handling. The results obtained are the same and the workflows are available on GitHub[24] and reusable.

---

24. Available at: https://github.com/fredericlemoine/workflows_bioinfo_iste.

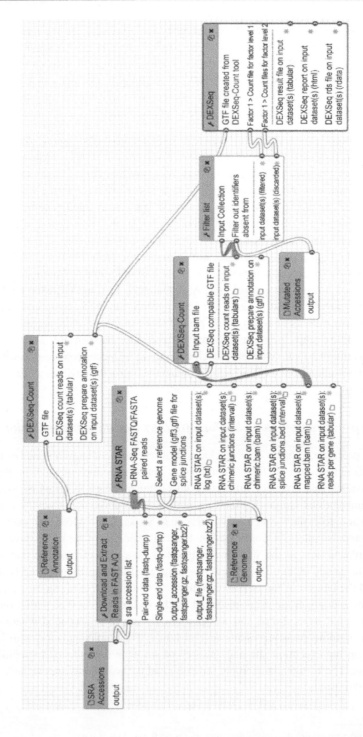

**Figure 3.5.** *Representation of the specification of the workflow by Galaxy (graphical interface)*

```
executor {
   name = 'slurm'
   queueSize = 2000
   queue = 'common'
   errorStrategy='retry'
   maxErrors=200
   maxRetries=200
}
process{
       executor='slurm'

       withName: 'getChromosomes' {
              executor='local'
       }
       withName: 'createGenomeIndex'{
              container='docker://flemoine/star'
              memory='30 GB'
              cpus=12
       }
       ...
}
```

**Figure 3.6.** *Extract from the Nextflow workflow configuration file*

## 3.5. Challenges, open problems and research opportunities

This section discusses the main challenges in using scientific workflows for automating data processing chains.

### 3.5.1. *Formalizing workflow development*

Many aspects of the tasks of implementing data processing chains are greatly simplified when a workflow system is used. The aspects related to the parameterization of execution sites (cloud, cluster, etc.), the management of dependencies related to the runtime environment and the management of access to data are largely the responsibility of the workflow system. Nevertheless, the development of complex data processing chains involving massive data still comprises many areas for improvement when workflow-based solutions are implemented. In this section, we list the main problems that remain to be solved, related to workflow development, from its creation, implying necessary testing phases, up to its maintenance.

### 3.5.2. Workflow testing

In software engineering, the testing task (unit, integration) is an integral part of the development process. In the context of multi-language scripts, especially those present in workflows, there is no systematic tool for testing and validating them. The challenge here is therefore to transpose classical software development practices (recently popularized with initiatives such as Software Carpentry by Wilson (2013)) to the entire lifecycle of a scientific workflow development process.

However, the software approach must take into account specific needs inherent to the bioinformatics domain. For example, the fact of providing sample data is part of many recommendations in workflow development: in the workflow4ever project (Hettne et al. 2012), the workflow designer should "provide sample inputs and outputs" (rule 4) and "test and validate" (rule 9); sharing repositories such as BioSharing have the same kind of expectations. Therefore, sharing platforms should allow the developers to upload datasets along the workflows. Nevertheless, there is the intrinsically difficult problem of providing, from the huge volumes of data on which the experiment has been run, small automatically generated datasets that are both statistically representative and do not pose a confidentiality problem.

#### 3.5.2.1. Workflow maintenance

Due to the evolving nature of computational experiments, the maintenance of data processing chains (implemented as scripts or workflows) is an important point. As a matter of fact, with their dependencies, they must be regularly updated, and this for different reasons. It is necessary to take into account the constant evolution of tools in order to take advantage of the improvements made, or to benefit from bug fixes or security problems. Virtualization techniques are particularly well suited for performing identical repetitions of experiments, but they do not allow for managing these types of updates.

Upgrading a virtual machine is as complex as upgrading a real machine. Version compatibility between different tools must be checked manually without assistance to automatically select the most recent consistent set of dependencies. It is necessary to have access to algorithms capable of determining as efficiently as possible the update of an environment according to a set of incompatibilities between packages. Solving this type of problem requires a mix of combinatorial algorithms and software engineering approaches.

#### 3.5.2.2. From workflows to scripts... and vice versa

In more general terms, it is the relationship between workflows and scripts that needs to be studied more closely. Because of the extremely high use of tools, closing the gap between the use of scripts and workflows to implement a multi-language data processing chain is of paramount importance and would have a considerable impact. Projects such as NoWorkflow (Chirigati et al. 2015) and YesWorkflow (McPhillips et al. 2015) have started to propose elements of solutions (such as the notion of the provenance of a script). However, more generally, it is the very way of thinking about development that needs to be defined with the different communities.

### 3.5.3. *Discovering and sharing workflows*

While many workflows are available today in workflow repositories (as myExperiment, for example), their reuse remains limited as it can be seen in several studies (Starlinger et al. 2012). Several avenues can be explored to address this point: (1) better documentation and annotation of the available workflows; (2) a more concise representation of the workflows to better understand them; (3) the exploitation of similarity between workflows to find close workflows; (4) the implementation of mechanisms for citing workflows; and (5) access to a more refined analysis of the results obtained by the workflows by exploiting provenance data.

#### 3.5.3.1. *Semi-automatic workflow annotation*

Several term ontologies describing bioinformatics data and tools have been designed (such as EDAM, for example), but there is currently no approach capable of helping users annotate their workflows using these terms. The challenge consists of designing text extraction (Hirschman et al. 2012) and data mining tools that can make the most of the various aspects of the workflow metadata, including the tools used in the workflow, the textual descriptions of the workflow but also the profile of the workflow designer.

#### 3.5.3.2. *Similarity between workflows*

The ability to discover alternative workflows that can implement the same analysis as the original workflow is of paramount importance. The definition of similarity in scientific workflows is central to this issue, and has been widely studied in recent years and is well understood (Cohen-Boulakia and Leser 2011; Bergmann and Gil 2014; Starlinger et al. 2014; Ma et al. 2015). A few new efficient algorithms have been designed to find workflows based on similarity in warehouses (Starlinger et al. 2016) but none have been deployed on workflow sharing platforms (yet).

#### 3.5.3.3. *Reducing workflow complexity*

When a workflow gets too complex (e.g. because it contains large numbers of data conversion steps), it becomes hardly comprehensible to the end-user.

Several solutions have been proposed to reduce the complexity of workflows. The Zoom system (Biton et al. 2007) reduces the workflow by automatically building sub-workflows based on the modules of interest specified by the user. DistillFlow (Cohen-Boulakia et al. 2014) searches for antipatterns in the workflow (redundancies), which it removes by exploiting rewriting rules. Other approaches (Alper et al. 2013; Gaignard et al. 2014) make use of the semantic annotations of the workflow modules to propose a simplified version. These approaches endeavor to reach the same objective but use very different techniques, from graph algorithms to the Semantic Web, thus illustrating the wide range of domains that can be used to propose solutions to the problem of reducing workflow complexity.

#### 3.5.3.4. Workflow citation

Similarly to scientific articles, datasets or software, several initiatives have come to the same conclusion: workflows, which encapsulate a given bioinformatics methodology, must be **citable objects**. Citations are then seen as a way to encourage sharing. The problem of citing workflows is different from the (much studied, see (Freire et al. 2012)) problem of tracing workflow history. The difficulty of this problem is related to the fact that the workflows to be considered can be (1) very complex graphs, (2) composed of nested workflows and (3) resulting from a combination of several reused workflows.

#### 3.5.3.5. Better use of provenance

Provenance (trace of the workflow runtime) is at the heart of the reproducibility problem, but also and above all of the problem of helping to understand the results. It is based on provenance-related information that the bioinformatician can understand and analyze different results from one execution to another when the parameters or data sets vary. More systematic adoption of workflow systems by the bioinformatics community will be facilitated if interactive tools capable of visualizing, querying or extracting provenance information are developed.

## 3.6. Conclusion

Numerous bioinformatics data processing chains are daily developed for analyzing massive data. These chains involve very heterogeneous types of tools (calls to Web services, R scripts, Python scripts, etc.) that need to be chained together and executed in optimized environments (clusters, clouds, etc.). The level of IT expertise required for this type of development is particularly high, especially when it comes to ensuring the reproducibility of the analysis over time. Workflow systems have been developed to guide the user in the development of these data processing chains. The use of these systems makes it possible to reduce the level of expertise required, when correctly utilized. In particular, workflow systems are characterized by three levels: specification, execution and runtime environment. These systems are based on strong characteristics: encapsulation and modularity (the execution of a workflow step should not depend on the execution of the following steps), independence, optimization and traceability (the system takes care of the scheduling and the interactions with the execution site). These main principles allow workflow developers to avoid many implementation difficulties. A more systematic adoption of this new programming paradigm depends on advances in many areas.

Finally, bioinformatics is fully in line with recent general principles of reproducible science, including the need to generate FAIR (Findable, Accessible, Interoperable and Reproducible) data. It is important to note that scientific workflows inherently contribute to the principles of FAIR data by processing data according to established metadata, creating metadata themselves when processing data and tracking and recording the provenance of data. These properties greatly facilitate the evaluation of data quality and contribute to data reuse. Since workflows are digital objects on their own, FAIR principles for workflows need to be defined to take into account their specific nature in terms of composition of executable software steps, their provenance and their development.

## 3.7. References

Alper, P., Belhajjame, K., Goble, C.A., Karagoz, P. (2013). Enhancing and abstracting scientific workflow provenance for data publishing. In *Joint 2013 EDBT/ICDT Conferences, EDBT/ICDT '13*, March 22, Genoa.

Amstutz, P., Andeer, R., Chapman, B., Chilton, J., Crusoe, M.R., Guimerà, R.V., Hernandez, G.C., Ivkovic, S., Kartashov, A., Kern, J. et al. (2016). Common workflow language (CWL) workflow description, draft 3. Common Workflow Language Working Group.

Anders, S., Reyes, A., Huber, W. (2012). Detecting differential usage of exons from RNA-seq data. *Genome Research*, 22(10), 2008–2017.

Belhajjame, K., Zhao, J., Garijo, D., Gamble, M., Hettne, K.M., Palma, R., Mina, E., Corcho, Ó., Gómez-Pérez, J.M., Bechhofer, S. et al. (2015). Using a suite of ontologies for preserving workflow-centric research objects. *J. Web Sem.*, 32, 16–42.

Bergmann, R. and Gil, Y. (2014). Similarity assessment and efficient retrieval of semantic workflows. *Information Systems*, 40, 115–127.

Berthold, M.R., Cebron, N., Dill, F., Gabriel, T.R., Kötter, T., Meinl, T., Ohl, P., Thiel, K., Wiswedel, B. (2009). Knime-the konstanz information miner: Version 2.0 and beyond. *ACM SIGKDD Explorations Newsletter*, 11(1), 26–31.

Bhagat, J., Tanoh, F., Nzuobontane, E., Laurent, T., Orlowski, J., Roos, M., Wolstencroft, K., Aleksejevs, S., Stevens, R., Pettifer, S. et al. (2010). Biocatalogue: A universal catalogue of web services for the life sciences. *Nucleic Acids Research*, 38(Suppl. 2), W689–W694.

Biton, O., Cohen-Boulakia, S., Davidson, S.B. (2007), Zoom*userviews: Querying relevant provenance in workflow systems. In *Proceedings of the 33rd International Conference on Very Large Data Bases*, University of Vienna, ACM, September 23–27.

Blankenberg, D., Von Kuster, G., Bouvier, E., Baker, D., Afgan, E., Stoler, N., Taylor, J., Nekrutenko, A. (2014). Dissemination of scientific software with Galaxy ToolShed. *Genome Biol.*, 15(2), 403.

Boettiger, C. (2015). An introduction to docker for reproducible research. *ACM SIGOPS Operating Systems Review*, 49(1), 71–79.

Cerezo, N. and Montagnat, J. (2011). Scientific workflow reuse through conceptual workflows on the virtual imaging platform. In *Proceedings of the 6th Workshop on Workflows in Support of Large-scale Science, WORKS '11*. ACM, New York.

Chirigati, F., Shasha, D., Freire, J. (2013). Reprozip: Using provenance to support computational reproducibility. In *International Workshop on Theory and Practice of Provenance*. TaPP, Lombard, Illinois.

Chirigati, F., Koop, D., Freire, J. (2015). Noworkflow: Capturing and analyzing provenance of scripts. In *Provenance and Annotation of Data and Processes: 5th International Provenance and Annotation Workshop, IPAW 2014*, June 9–13, 2014. Revised Selected Papers, volume 8628. Springer, Cologne.

Cock, P.J.A., Antao, T., Chang, J.T., Chapman, B.A., Cox, C.J., Dalke, A., Friedberg, I., Hamelryck, T., Kauff, F., Wilczynski, B., de Hoon, M.J.L. (2009). Biopython: Freely available Python tools for computational molecular biology and bioinformatics. *Bioinformatics*, 25(11), 1422–1423.

Cohen-Boulakia, S. and Leser, U. (2011). Search, adapt, and reuse: The future of scientific workflows. *ACM SIGMOD Record*, 40(2), 6–16.

Cohen-Boulakia, S., Chen, J., Missier, P., Goble, C., Williams, A., Froidevaux, C. (2014). Distilling structure in Taverna scientific workflows: A refactoring approach. *BMC Bioinformatics*, 15(Suppl. 1), S12.

Costa, F., Silva, V., de Oliveira, D., Ocaña, K.A.C.S., Ogasawara, E.S., Dias, J., Mattoso, M. (2013). Capturing and querying workflow runtime provenance with PROV: A practical approach. In *Joint 2013 EDBT/ICDT Conferences, EDBT/ICDT '13*, March 22. Genoa.

Di Tommaso, P., Chatzou, M., Floden, E.W., Barja, P.P., Palumbo, E., Notredame, C. (2017). Nextflow enables reproducible computational workflows. *Nature Biotechnology*, 35(4), 316.

Dobin, A., Davis, C.A., Schlesinger, F., Drenkow, J., Zaleski, C., Jha, S., Batut, P., Chaisson, M., Gingeras, T.R. (2013). Star: Ultrafast universal RNA-seq aligner. *Bioinformatics*, 29(1), 15–21.

Freire, J., Bonnet, P., Shasha, D. (2012). Computational reproducibility: State-of-the-art, challenges, and database research opportunities. In *Proceedings of the 2012 ACM SIGMOD International Conference on Management of Data*. ACM.

Furlani, J.L. (1991). Modules: Providing a flexible user environment. In *Proceedings of the Fifth Large Installation Systems Administration Conference (LISA V)*.

Furney, S.J., Pedersen, M., Gentien, D., Dumont, A.G., Rapinat, A., Desjardins, L., Turajlic, S., Piperno-Neumann, S., de la Grange, P., Roman-Roman, S. et al. (2013). Sf3b1 mutations are associated with alternative splicing in uveal melanoma. *Cancer Discovery*, 3(10), 1122–1129.

Gaignard, A., Montagnat, J., Gibaud, B., Forestier, G., Glatard, T. (2014). Domain-specific summarization of life-science e-experiments from provenance traces. *Web Semantics: Science, Services and Agents on the World Wide Web*, 29, 19–30.

Giardine, B., Riemer, C., Hardison, R.C., Burhans, R., Elnitski, L., Shah, P., Zhang, Y., Blankenberg, D., Albert, I., Taylor, J. et al. (2005). Galaxy: A platform for interactive large-scale genome analysis. *Genome Research*, 15(10), 1451–1455.

Gil, Y., Ratnakar, V., Kim, J., Gonzalez-Calero, P., Groth, P., Moody, J., Deelman, E. (2010). Wings: Intelligent workflow-based design of computational experiments. *IEEE Intelligent Systems*, 26(1), 62–72.

Goecks, J., Nekrutenko, A., Taylor, J., The Galaxy Team (2010). Galaxy: A comprehensive approach for supporting accessible, reproducible, and transparent computational research in the life sciences. *Genome Biol.*, 11(8), R86.

Gonzalez-Beltran, A., Li, P., Zhao, J., Avila-Garcia, M.S., Roos, M., Thompson, M., van der Horst, E., Kaliyaperumal, R., Luo, R., Lee, T.-L. et al. (2014). From peer-reviewed to peer-reproduced: A role for data standards, models and computational workflows in scholarly publishing. *bioRxiv*.

Guo, P. (2012). CDE: A tool for creating portable experimental software packages. *Computing in Science & Engineering*, 14(4), 32–35.

Harbour, J.W., Roberson, E.D., Anbunathan, H., Onken, M.D., Worley, L.A., Bowcock, A.M. (2013). Recurrent mutations at codon 625 of the splicing factor SF3B1 in uveal melanoma. *Nature Genetics*, 45(2), 133.

Hettne, K.M., Wolstencroft, K., Belhajjame, K., Goble, C.A., Mina, E., Dharuri, H., De Roure, D., Verdes-Montenegro, L., Garrido, J., Roos, M. (2012). Best practices for workflow design: How to prevent workflow decay. *SWAT4LS*.

Hirschman, L., Burns, G.A.C., Krallinger, M., Arighi, C., Cohen, K.B., Valencia, A., Wu, C.H., Chatr-Aryamontri, A., Dowell, K.G., Huala, E. et al. (2012). Text mining for the biocuration workflow. *Database*, 2012, bas020.

Ison, J.E.A. (2013). Edam: An ontology of bioinformatics operations, types of data and identifiers, topics and formats. *Bioinformatics*, 29(10), 1325–1332.

Ison, J.E.A., Rapacki, K., Ménager, H., Kalaš, M., Rydza, E., Chmura, P., Anthon, C., Beard, N., Berka, K., Bolser, D. et al. (2016). Tools and data services registry: A community effort to document bioinformatics resources. *Nucleic Acids Research*, 44(D1), D38–D47.

Jette, M.A., Yoo, A.B., Grondona, M. (2002). Slurm: Simple linux utility for resource management. In *Lecture Notes in Computer Science: Proceedings of Job Scheduling Strategies for Parallel Processing (JSSPP) 2003*. Springer-Verlag.

Juve, G., Chervenak, A.L., Deelman, E., Bharathi, S., Mehta, G., Vahi, K. (2013). Characterizing and profiling scientific workflows. *Future Generation Comp. Syst.*, 29(3), 682–692.

Koenig, D., Glover, A., King, P., Laforge, G., Skeet, J. (2007). *Groovy in Action*, Volume 1. Manning, New York.

Korkhov, V., Krefting, D., Montagnat, J., Huu, T.T., Kukla, T., Terstyanszky, G., Manset, D., Caan, M., Olabarriaga, S. (2012). Shiwa workflow interoperability solutions for neuroimaging data analysis. *Stud. Health Technol. Inform.*, 175, 109–110.

Köster, J. and Rahmann, S. (2012). Snakemake – A scalable bioinformatics workflow engine. *Bioinformatics*, 28(19), 2520–2522.

Kurtzer, G.M., Sochat, V., Bauer, M.W. (2017). Singularity: Scientific containers for mobility of compute. *PLoS One*, 12(5), e0177459.

Li, H., Handsaker, B., Wysoker, A., Fennell, T., Ruan, J., Homer, N., Marth, G., Abecasis, G., Durbin, R. (2009). The sequence alignment/map format and samtools. *Bioinformatics*, 25(16), 2078–2079.

Ludäscher, B., Altintas, I., Berkley, C., Higgins, D., Jaeger, E., Jones, M., Lee, E.A., Tao, J., Zhao, Y. (2006). Scientific workflow management and the Kepler system. *Concurrency and Computation: Practice and Experience*, 18(10), 1039–1065.

Ma, Y., Shi, M., Wei, J. (2015). Cost and accuracy aware scientific workflow retrieval based on distance measure. *Information Sciences*, 314, 1–13.

Mates, P., Santos, E., Freire, J., Silva, C.T. (2011). Crowdlabs: Social analysis and visualization for the sciences. *Scientific and Statistical Database Management*, 555–564.

McPhillips, T., Song, T., Kolisnik, T., Aulenbach, S., Belhajjame, K., Bocinsky, K., Cao, Y., Chirigati, F., Dey, S., Freire, J. et al. (2015). Yesworkflow: A user-oriented, language-independent tool for recovering workflow information from scripts. *arXiv preprint arXiv:1502.02403*.

Plankensteiner, K., Montagnat, J., Prodan, R. (2011). IWIR: A language enabling portability across grid workflow systems. In *Proceedings of the 6th Workshop on Workflows in Support of Large-scale Science, WORKS '11*. ACM, New York.

Pradal, C., Artzet, S., Chopard, J., Dupuis, D., Fournier, C., Mielewczik, M., Nègre, V., Neveu, P., Parigot, D., Valduriez, P. et al. (2017). Infraphenogrid: A scientific workflow infrastructure for plant phenomics on the grid. *Future Generation Comp. Syst.*, 67, 341–353.

Roure, D.D., Goble, C.A., Stevens, R. (2009). The design and realisation of the myExperiment Virtual Research Environment for social sharing of workflows. *Future Generation Comp. Syst.*, 25(5), 561–567.

Sandve, G.K., Nekrutenko, A., Taylor, J., Hovig, E. (2013). Ten simple rules for reproducible computational research. *PLoS Computational Biology*, 9(10), 1–4.

Starlinger, J., Cohen-Boulakia, S., Leser, U. (2012). (Re)use in public scientific workflow repositories. In *Scientific and Statistical Database Management – 24th International Conference, SSDBM 2012*, June 25–27. Chania, Crete.

Starlinger, J., Brancotte, B., Cohen-Boulakia, S., Leser, U. (2014). Similarity search for scientific workflows. *Proceedings of the VLDB Endowment*, 7(12), 1143–1154.

Starlinger, J., Cohen-Boulakia, S., Khanna, S., Davidson, S.B., Leser, U. (2016). Effective and efficient similarity search in scientific workflow repositories. *Future Generation Computer Systems*, 56, 584–594.

Wilson, G. (2013). Software carpentry: Lessons learned. *F1000Research*, 3, 62–62.

Wolstencroft, K., Haines, R., Fellows, D., Williams, A., Withers, D., Owen, S., Soiland-Reyes, S., Dunlop, I., Nenadic, A., Fisher, P. et al. (2013). The taverna workflow suite: Designing and executing workflows of web services on the desktop, web or in the cloud. *Nucleic Acids Research*, 41(Web server issue), W557–W561.

# PART 2

# Integration and Statistics

# 4

# Variable Selection in the General Linear Model: Application to Multiomic Approaches for the Study of Seed Quality

Céline LÉVY-LEDUC[1], Marie PERROT-DOCKÈS[2], Gwendal CUEFF[3] and Loïc RAJJOU[3]

[1] Université Paris-Saclay, AgroParisTech, INRAE, MIA Paris-Saclay, France
[2] Université Paris Cité, MAP5, CNRS, France
[3] Université Paris-Saclay, INRAE, AgroParisTech, IJPB, Versailles, France

Omics data are characterized by the presence of a strong dependence structure that often results from biological processes. The application of variable selection methods that do not take this underlying dependence structure into account can lead to the selection of irrelevant variables. The aim of this chapter is to propose a new method of variable selection in the general linear model that takes into account the dependence that may exist between the columns of the observation matrix, which corresponds to the variables. We show that the fact of including the estimation of the covariance matrix of the observations in the Lasso (least absolute shrinkage and selection operator) criterion significantly improves the performance of variable

For a color version of all figures in this chapter, see: http://www.iste.co.uk/froidevaux/biologicaldata.zip.

*Biological Data Integration*,
coordinated by Christine FROIDEVAUX, Marie-Laure MARTIN-MAGNIETTE and Guillem RIGAILL © ISTE Ltd 2023.

selection. We have implemented our method in the `MultiVarSel` R package, which is available through the CRAN (Comprehensive R Archive Network), and applied it to omics data to study seed quality.

## 4.1. Introduction

Sexual reproduction of flowering plants leads to seed formation, which is the primary driver for the dispersion and propagation of plant species. Seeds are reserve organs, rich in proteins, lipids or carbohydrates, which are necessary for germination and seedling emergence. They are widely used by humans for both food and non-food uses. The environmental conditions to which the mother plants are subjected have a strong impact on yields and the quality of the seeds produced. This quality is developed in the field and refined at harvest time (sorting, processing), maintained in storage and expressed at the time of sowing. The quality can be expressed by different characteristics such as filling rate, morphology, storage capacity or the germination potential (Blödner et al. 2007; Burghardt et al. 2016). This last point is crucial for farmers. A good quality batch of seeds in agricultural production must have a good germination vigor, in other words, seeds that will germinate quickly and evenly and therefore with very low dormancy. This homogeneous germination makes it possible to synchronize the whole crop development cycle, field emergence, plant growth, flowering and harvesting dates. In *Arabidopsis thaliana*, the model plant, the seed production temperature affects their potential for germination (Springthorpe and Penfield 2015). Low temperatures during seed maturation stimulate dormancy. Also, in the present study, *Arabidopsis* seeds were produced at a low temperature (14–16°C), at an intermediate temperature (18–22°C) and at a high temperature (25–28°C), offering biological material of variable quality. Diagnosing seed quality is a real challenge for the industry. Analyses of gene expression products (transcripts, proteins) and biochemical composition (metabolites) using high-throughput "omics"-based approaches (transcriptomics, proteomics and metabolomics) offer a strong potential for characterizing seed quality biomarkers. A number of experiments have shown a strong disagreement between transcript accumulation and the abundance of corresponding proteins in the seeds (Galland et al. 2014). This is corroborated by previous observations indicating that transcript abundance does not necessarily reflect its translation into protein, especially under stressful conditions (Bailey-Serres et al. 2009). Here, we have focused our search for biomarkers on proteomic and metabolomic analyses of dry freshly harvested mature seeds.

In order to understand the effect of temperature on the response of a given metabolite or protein observed with $n$ independent samples, the one-way analysis of variance (ANOVA) analysis model can be used as follows:

$$\begin{pmatrix} Y_{1,1} \\ Y_{2,1} \\ \vdots \\ Y_{n,1} \end{pmatrix} = \begin{pmatrix} 1 & 0 & 0 \\ \vdots & \vdots & \vdots \\ 1 & 0 & 0 \\ 0 & 1 & 0 \\ \vdots & \vdots & \vdots \\ 0 & 1 & 0 \\ 0 & 0 & 1 \\ \vdots & \vdots & \vdots \\ 0 & 0 & 1 \end{pmatrix} \begin{pmatrix} B_{1,1} \\ B_{2,1} \\ B_{3,1} \end{pmatrix} + \begin{pmatrix} E_{1,1} \\ E_{2,1} \\ \vdots \\ E_{n,1} \end{pmatrix}$$

where $E_{i,1}$ are assumed to be independent-centered Gaussian random variables for the purpose of estimating the coefficients $B_{j,1}$. When $\widehat{B}_{j,1} > 0$ (respectively, $\widehat{B}_{j,1} < 0$), this would mean that metabolite 1 or protein 1 tends to be over-accumulated (respectively, under-accumulated) in the modality $j$ of the temperature variable. During a metabolomics or proteomics experiment, we generally have access to the responses of $q$ metabolites or proteins. In a metabolomics experiment, one is led, for example, to consider the following MANOVA (multivariate ANOVA) model:

$$Y = XB + E \qquad [4.1]$$

where:

$$\underset{n \times q}{Y} = \begin{pmatrix} \text{Metabolite 1} & \text{Metabolite 2} & \cdots & \text{Metabolite } q \\ Y_{1,1} & Y_{1,2} & \cdots & Y_{1,q} \\ Y_{2,1} & Y_{2,2} & \cdots & Y_{2,q} \\ \vdots & \vdots & & \vdots \\ Y_{n,1} & Y_{n,2} & \cdots & Y_{n,q} \end{pmatrix}, \underset{n \times p}{X} = \begin{pmatrix} 1 & 0 & 0 \\ \vdots & \vdots & \vdots \\ 1 & 0 & 0 \\ 0 & 1 & 0 \\ \vdots & \vdots & \vdots \\ 0 & 1 & 0 \\ 0 & 0 & 1 \\ \vdots & \vdots & \vdots \\ 0 & 0 & 1 \end{pmatrix}$$

$$\underset{p\times q}{\boldsymbol{B}} = \begin{pmatrix} B_{1,1} & B_{1,2} & \cdots & B_{1,q} \\ B_{2,1} & B_{2,2} & \cdots & B_{2,q} \\ B_{p,1} & B_{p,2} & \cdots & B_{p,q} \end{pmatrix} \text{ and } \underset{n\times q}{\boldsymbol{E}} = \begin{pmatrix} E_{1,1} & E_{1,2} & \cdots & E_{1,q} \\ E_{2,1} & E_{2,2} & \cdots & E_{2,q} \\ \vdots & \vdots & & \vdots \\ E_{n,1} & E_{n,2} & \cdots & E_{n,q} \end{pmatrix}$$

For a more detailed presentation of the MANOVA model and its inference, we refer the reader to Mardia et al. (1979) and Muller and Stewart (2006). This model is part of the general linear models that should not be confused with generalized linear models. In this case, $(Y_{1,k}, \ldots, Y_{n,k})'$ represents the response of metabolite $k$ on $n$ samples and by estimating the $B_{j,k}$, it will be possible to know if the modality $j$ of the temperature variable has a positive, negative or null effect on the metabolite $k$.

Different statistical learning approaches have been proposed to analyze "omics" data; they are described in some previous studies (Saccenti et al. 2013; Ren et al. 2015; Boccard and Rudaz 2016; Zhang et al. 2017). Among them, PLS-DA and sPLS-DA proposed in Lê Cao et al. (2011); Durif et al. (2017) and el Bouhaddani et al. (2018) can be mentioned.

Assuming that each column of the observation matrix $\boldsymbol{Y}$ has a zero empirical mean, the question of understanding whether a modality of the temperature variable has an influence or not on a metabolite or a protein is a variable selection problem in the general linear model (see section 4.1). Several approaches can be used to do variable selection in this type of model. Conventional statistical tests can be used in univariate ANOVA models such as those described in Mardia et al. (1979) or Faraway (2004) to analyze each column of $\boldsymbol{Y}$ separately. On each column of $\boldsymbol{Y}$, Lasso-based approaches such as those described in Tibshirani (1996) can also be utilized. However, these methods do not take into account the potential dependence that may exist between the columns of $\boldsymbol{Y}$.

In this chapter, we propose a method that models the dependence that can exist between the columns $\boldsymbol{Y}$, and uses it to select variables. More precisely, we will assume that the rows $\boldsymbol{E}$ are independent and that for each row $i$, the vector $\boldsymbol{E}_i$ is a Gaussian vector with zero expectation and covariance matrix $\boldsymbol{\Sigma}_q$:

$$\boldsymbol{E}_i = (E_{i,1}, \ldots, E_{i,q}) \sim \mathcal{N}(0, \boldsymbol{\Sigma}_q) \qquad [4.2]$$

Properly estimating $\boldsymbol{\Sigma}_q$ is in general impossible in the case where $n << q$ without additional assumptions. We will assume here that each $\boldsymbol{E}_i$ can be modeled as a realization of a stationary process and we will show how $\boldsymbol{\Sigma}_q$ can be estimated within this framework. It should be noted that our method removes the potential dependence between columns and then employs a Lasso approach associated with a *stability selection* step proposed by Meinshausen and Buhlmann (2010) to make sure that only the most stable variables are kept. Our method is implemented in the R package MultiVarSel, which is available through the CRAN.

The remaining chapter is organized as follows. Our methodology is described in section 4.2. We propose to validate it from numerical simulations in section 4.3, and we apply it to metabolomics and proteomics data in section 4.4.

## 4.2. Methodology

The method that we propose can be summarized as follows:

– step 1: assuming that each column of the matrix $Y$ follows a one-way ANOVA model, we obtain an estimate $\widehat{E}$ of the error matrix $E$;

– step 2: the matrix $\Sigma_q$ is estimated using the methods described in sections 4.2.1.1 and 4.2.1.2. Then, the estimator $\widehat{\Sigma}_q$ of $\Sigma_q$ is chosen as the most suitable based on the statistical test described in section 4.2.1.3;

– step 3: using $\widehat{\Sigma}_q$, we transform the data in order to remove the dependence existing between the columns of the matrix $Y$;

– step 4: the Lasso method described in section 4.2.2 is applied to the transformed data.

The first step provides a preliminary estimator $\widetilde{B}$ of $B$. An estimator $\widehat{E}$ of $E$ is then defined by:

$$\widehat{E} = Y - X\widetilde{B} \qquad [4.3]$$

In the following, we will focus on the description of the other three steps.

### 4.2.1. *Estimation of the covariance matrix* $\Sigma_q$

In the following, in order to estimate $\Sigma_q$, we propose to model each row of $E$ as the realization of a stationary process, we will therefore use several types of stationary processes such as the ARMA processes, for example. For more details on this subject, the reader may consult the book by Brockwell and Davis (1991).

In this section, we will consider different models of stationary processes, and special attention will be given to the estimation of $\Sigma_q^{-1/2}$ since we will use the following transformation:

$$Y \Sigma_q^{-1/2} = XB \Sigma_q^{-1/2} + E \Sigma_q^{-1/2} \qquad [4.4]$$

to remove the dependence between the columns of $\mathbf{Y}$. Actually, the correlation matrix of each row of $\mathbf{E}\Sigma_q^{-1/2}$ is equal to the identity matrix. Such a procedure will be called "whitening" in the remainder of the chapter.

### 4.2.1.1. *ARMA process*

One of the simplest ARMA processes is the first-order autoregressive process of (AR(1)). More precisely, this amounts to assuming that for each $i$ in $\{1,\ldots,n\}$, $E_{i,t}$ verifies the following equation:

$$E_{i,t} - \phi_1 E_{i,t-1} = W_{i,t}, \forall t \in \mathbb{Z} \qquad [4.5]$$

where $|\phi_1| < 1$ and $W_{i,t} \sim BB(0,\sigma^2)$. The notation $BB(0,\sigma^2)$ denotes white noise with zero expectation and variance $\sigma^2$, defined as follows:

$$Z_t \sim BB(0,\sigma^2) \text{ if } \begin{cases} \mathbb{E}(Z_t) = 0, \\ \mathbb{E}(Z_t Z_{t'}) = 0 \text{ if } t \neq t' \\ \mathbb{E}(Z_t^2) = \sigma^2 \end{cases} \qquad [4.6]$$

It should be noted that the closer the parameter $\phi_1$ is to 1, the greater the dependence between the $E_{i,t}$ at $i$.

When $\sigma^2 = 1$, the inverse of the square root of $\Sigma_q$ assumes the following simple explicit form:

$$\Sigma_q^{-1/2} = \begin{pmatrix} \sqrt{1-\phi_1^2} & -\phi_1 & 0 & \cdots & 0 \\ 0 & 1 & -\phi_1 & \cdots & 0 \\ 0 & 0 & \ddots & \ddots & \vdots \\ \vdots & \vdots & \ddots & \ddots & -\phi_1 \\ 0 & 0 & \cdots & 0 & 1 \end{pmatrix} \qquad [4.7]$$

It should be underlined that when $\sigma^2$ is not equal to 1, the covariance matrix of each row of $\mathbf{E}\Sigma_q^{-1/2}$ is equal to $\sigma^2 \text{Id}$ and that the correlation matrix is equal to the identity matrix.

Therefore, to obtain an estimator $\widehat{\Sigma}_q^{-1/2}$ of $\Sigma_q^{-1/2}$, we simply have to know how to estimate the parameter $\phi_1$ and replace it by its estimator $\widehat{\phi}_1$ in equation [4.7]. To this end, we use $\widehat{E}$ defined by equation [4.3] and define $\widehat{\phi}_1$ as:

$$\widehat{\phi}_1 = \frac{1}{n}\sum_{i=1}^{n}\widehat{\phi}_{1,i}$$

where $\widehat{\phi}_{1,i}$ denotes the estimator of $\phi_1$ obtained by the Yule–Walker equations from $(\widehat{E}_{i,1},\ldots,\widehat{E}_{i,q})$; see Brockwell and Davis (1991) for more information about this method.

In general, it is also possible to have access to $\Sigma_q^{-1/2}$ for processes ARMA$(p,q)$ defined as follows, for each $i$ in $\{1,\ldots,n\}$:

$$E_{i,t} - \phi_1 E_{i,t-1} - \cdots - \phi_p E_{i,t-p} = W_{i,t} + \theta_1 W_{i,t-1} + \ldots \theta_q W_{i,t-q} \quad [4.8]$$

where $W_{i,t} \sim BB(0,\sigma^2)$, $\phi_k$ and $\theta_k$ are real parameters.

#### 4.2.1.2. *General stationary processes*

In problems where modeling by an ARMA$(p,q)$ process is not appropriate, each row of $E$ can be modeled as a general weakly stationary process and $\Sigma_q$ estimated as follows:

$$\widehat{\Sigma}_q = \begin{pmatrix} \widehat{\gamma}(0) & \widehat{\gamma}(1) & \cdots & \widehat{\gamma}(q-1) \\ \widehat{\gamma}(1) & \widehat{\gamma}(0) & \cdots & \widehat{\gamma}(q-2) \\ \vdots & & & \\ \widehat{\gamma}(q-1) & \widehat{\gamma}(q-2) & \cdots & \widehat{\gamma}(0) \end{pmatrix} \quad [4.9]$$

where:

$$\widehat{\gamma}(h) = \frac{1}{n}\sum_{i=1}^{n} \widehat{\gamma}_i(h)$$

and $\widehat{\gamma}_i(h)$ is the classical estimator of $\gamma_i(h) = \mathbb{E}(E_{i,t}E_{i,t+h})$ for all $t$. More precisely, $\widehat{\gamma}_i(h)$ is the empirical autocovariance of $\widehat{E}_{i,t}$ in $h$, that is to say the empirical covariance between $(\widehat{E}_{i,1},\ldots,\widehat{E}_{i,n-h})$ and $(\widehat{E}_{i,h+1},\ldots,\widehat{E}_{i,n})$. For more details, see Chapter 7 of Brockwell and Davis (1991). The matrix $\widehat{\Sigma}_q^{-1/2}$ is then obtained by inverting the Cholesky factor of $\widehat{\Sigma}_q$.

#### 4.2.1.3. *Choosing the best dependence structure*

In order to know which dependence model is the most adapted, we propose to use the statistical test defined hereafter. If the dependence structure is chosen correctly, each row of $\widetilde{E} = \widehat{E}\widehat{\Sigma}_q^{-1/2}$ must be white noise defined in equation [4.6], where $\widehat{E}$ is given by equation [4.3].

In order to test whether a random process is white noise, one of the most commonly used methods is the Portmanteau test, which is based on the Bartlett theorem (see theorem 7.2.2 in (Brockwell and Davis 1991)). According to this theorem, we have that under the null hypothesis $(H_0)$: "For each $i$ in $\{1,\ldots,n\}$, $(\widetilde{E}_{i,1},\ldots,\widetilde{E}_{i,q})$ is white noise":

$$q\sum_{h=1}^{H}\widehat{\rho}_i(h)^2 \approx \chi^2(H), \text{ when } q \to \infty \qquad [4.10]$$

for each $i$ in $\{1,\ldots,n\}$, where $\widehat{\rho}_i(h)$ denotes the empirical autocorrelation of $(\widetilde{E}_{i,1},\ldots,\widetilde{E}_{i,q})$ in $h$ and $\chi^2(H)$ denotes the chi-square distribution with $H$ degrees of freedom. Thereby, according to equation [4.10], we have a $p$-value for each $i$ in $\{1,\ldots,n\}$. In order to have a single $p$-value instead of $n$, we will consider the following approximation:

$$q\sum_{i=1}^{n}\sum_{h=1}^{H}\widehat{\rho}_i(h)^2 \approx \chi^2(nH), \text{ when } q \to \infty \qquad [4.11]$$

where the approximation derives from the fact that the rows $\widetilde{E}$ are assumed to be independent. Equation [4.11] thus provides a $p$-value: Pval. It can therefore be concluded that if Pval $\leq \alpha$, we reject $(H_0)$ at level $\alpha$, where $\alpha$ is in general equal to 5%. If on the contrary Pval $> \alpha$, it means that the dependence structure of $E$ is well chosen. If none of the different dependence structures tested are well chosen, it will be possible to test a block dependence structure such as the one proposed in Perrot-Dockès et al. (2019).

### 4.2.2. Estimation of $\mathcal{B}$

#### 4.2.2.1. Lasso approach

We first recall the context in which the Lasso approach is usually utilized. Let us consider the following univariate linear model:

$$\mathcal{Y} = \mathcal{X}\mathcal{B} + \mathcal{E} \qquad [4.12]$$

where $\mathcal{Y}$, $\mathcal{B}$ and $\mathcal{E}$ are vectors. In general, in high-dimensional linear models, matrix $\mathcal{X}$ has more columns than rows, which means that the number of variables is greater than the number of observations and $\mathcal{B}$ is in general a parsimonious vector, which means that it has many zero components.

In such models, the Lasso method (*Least Absolute Shrinkage and Selection Operator*) initially proposed by Tibshirani (1996) is widely used. It is defined as follows for $\lambda > 0$:

$$\widehat{\mathcal{B}}(\lambda) = \operatorname{Argmin}_\mathcal{B} \left\{ \|\mathcal{Y} - \mathcal{XB}\|_2^2 + \lambda \|\mathcal{B}\|_1 \right\} \quad [4.13]$$

where for $u = (u_1, \ldots, u_n)$, $\|u\|_2^2 = \sum_{i=1}^n u_i^2$ and $\|u\|_1 = \sum_{i=1}^n |u_i|$, which corresponds to the norm $\ell_1$ of vector $u$. In equation [4.13], the first term corresponds to the least squares criterion and $\lambda \|\mathcal{B}\|_1$ can be seen as a penalty. The interest of the Lasso approach resides in providing a parsimonious estimator $\widehat{\mathcal{B}}$ and $\mathcal{B}$. A small number of non-zero components of $\widehat{\mathcal{B}}$ implies that $\lambda$ is large.

This methodology cannot be directly applied to our model since the observations available are not vectors but matrices. Nonetheless, we will see that model [4.1] can be rewritten as equation [4.12], where $\mathcal{Y}$, $\mathcal{B}$ and $\mathcal{E}$ are vectors of sizes $nq$, $pq$ and $nq$, respectively. Actually, if $vec(A)$ denotes the vector obtained from the matrix $A$ by stacking the columns of $A$ beneath one another and the $vec$ operator is applied to model [4.1], it then follows that:

$$vec(Y) = vec(XB + E) = vec(XB) + vec(E)$$

If we set $\mathcal{Y} = vec(Y)$, $\mathcal{B} = vec(B)$ and $\mathcal{E} = vec(E)$, we obtain:

$$\mathcal{Y} = vec(XB) + \mathcal{E} = (I_q \otimes X)\mathcal{B} + \mathcal{E}$$

where the following identity was employed:

$$vec(AXB) = (B' \otimes A)vec(X)$$

according to Appendix A.2.5 in Mardia et al. (1979). In this equation, $B'$ designates the transpose of matrix $B$. Therefore:

$$\mathcal{Y} = \mathcal{XB} + \mathcal{E}$$

where $\mathcal{X} = I_q \otimes X$ and $\mathcal{Y}$, $\mathcal{B}$ and $\mathcal{E}$ are vectors of sizes $nq$, $pq$ and $nq$, respectively.

Estimating the positions of the non-zero components of $\mathcal{B}$ is thus a first approach to select the most relevant variables. Nevertheless, this method does not take into account the dependence that may potentially exist between the columns of $Y$. In the following, we propose a modified version of the standard Lasso method, taking into account this potential dependence.

As explained earlier, our method first involves "whitening" the observations, in other words, removing the dependence that may exist between the columns of the observation matrix by multiplying equation [4.1] on the right-hand side by $\widehat{\Sigma}_q^{-1/2}$ (see equation [4.4], where $\Sigma_q^{-1/2}$ is replaced by $\widehat{\Sigma}_q^{-1/2}$). Using the same vectorization device as above to transform model [4.1] into [4.12], the Lasso criterion can be applied to the vectorized version of model [4.4], where $\Sigma_q^{-1/2}$ is replaced by $\widehat{\Sigma}_q^{-1/2}$. The expressions for $\mathcal{Y}$, $\mathcal{X}$, $\mathcal{B}$ and $\mathcal{E}$ can be obtained as follows.

By applying the $vec$ operator to model [4.4] in which $\Sigma_q^{-1/2}$ is replaced by $\widehat{\Sigma}_q^{-1/2}$, we obtain:

$$vec(Y\widehat{\Sigma}_q^{-1/2}) = vec(XB\widehat{\Sigma}_q^{-1/2}) + vec(E\widehat{\Sigma}_q^{-1/2})$$
$$= ((\widehat{\Sigma}_q^{-1/2})' \otimes X)vec(B) + vec(E\widehat{\Sigma}_q^{-1/2})$$

Therefore:

$$\mathcal{Y} = \mathcal{XB} + \mathcal{E} \qquad [4.14]$$

where $\mathcal{Y} = vec(Y\widehat{\Sigma}_q^{-1/2})$, $\mathcal{X} = (\widehat{\Sigma}_q^{-1/2})' \otimes X$ and $\mathcal{E} = vec(E\widehat{\Sigma}_q^{-1/2})$.

It should be noted that Rothman et al. (2010) also had a similar idea about "whitening" that they implemented in the R package MRCE. However, we observed through numerical simulations that the values of $n$ and $q$ that we wanted to use for the computation time of their approach was so large for a given value of the parameter $\lambda$ that it was impossible to use their method in our context.

### 4.2.2.2. *Choosing the regularization parameter*

The estimator defined in equation [4.13] depends on parameter $\lambda$ that allows the calibration of the level of parsimony of $\widehat{B}$, that is, the proportion of zero components of this estimator. To adjust the level of parsimony of $\widehat{B}$, we propose to use two classical approaches: cross-validation and the *stability selection* that ensure the robustness of the selected variables.

More precisely, we start by applying our method to $\mathcal{Y}$ defined in equation [4.14] and we obtain $\lambda_{CV}$. We then randomly choose a subsample of size $nq/2$ of $\mathcal{Y}$, apply our method with $\lambda = \lambda_{CV}$ and record the indices $i$ of the non-zero components of

$\widehat{\mathcal{B}}(\lambda)$. We repeat these steps of random subsampling and application of the Lasso criterion $N$ times. At the end of these various stages, we have access to the number of times $N_i$, where each component $\widehat{\mathcal{B}}_i$ of $\widehat{\mathcal{B}}$ has been estimated to be non-zero. We keep the $i$ components whose frequency $N_i/N$ is greater than a given threshold. The impact of the choice of $N$ and of the threshold is studied in section 4.3.

It should be noted that in Perrot-Dockès et al. (2018), we have established theoretical sign consistency results for the estimator $\widehat{\mathcal{B}}$ to validate our approach.

## 4.3. Numerical experiments

The aim of this section is to evaluate the statistical and numerical performance of our methodology that we have implemented in the R package MultiVarSel and to compare it to existing methods.

For this purpose, we generated observations $Y$ verifying equation [4.1] with $q = 1,000$, $p = 3$, $n = 30$ ($n_1 = 9$, $n_2 = 8$, $n_3 = 13$, $n_k$ corresponding to the number of repetitions of the modality $k$ of the qualitative variable, that is, the number of 1s in column $k$ of matrix $X$) and different dependence structures, that is, different matrices $\Sigma_q$ associated with the AR(1) model described in equation [4.5] with $\sigma = 1$, $\phi_1 = 0.7$ or 0.9.

It should be underlined that the values of parameters $p$, $q$ and $n$ were chosen because they correspond to orders of magnitude generally used in metabolomics and proteomics.

In this section, we will also study the effect of parsimony and the signal-to-noise ratio (SNR) on the statistical performance of the methods. The parsimony level $s$ corresponds to the proportion of non-zero elements in $\mathcal{B}$ and different SNR are obtained by multiplying $B$ defined in equation [4.1] by a coefficient $\kappa$.

### 4.3.1. *Statistical performance*

The purpose of this section is to compare the performance of our different whitening methods to existing methods.

#### 4.3.1.1. *Performance in terms of variable selection*

We compare our approach to the ANOVA method, to the standard Lasso method, that is, the Lasso criterion without whitening step, and to the sPLSDA method which is widely used in the field of metabolomics and described in Lê Cao et al. (2011). It should also be noted that this method is implemented in the R package mixOmics and is available on the MetaboAnalyst website.

The ANOVA method consists of modeling each column of the matrix of observations $Y$ as a one-way ANOVA, in such a manner as if the columns of $Y$ were independent. Our different whitening methods described in sections 4.2.1.1 and 4.2.1.2 are denoted AR1 and Nonparam. They are compared to the Oracle method where $\Sigma_q$ is known, which is never the case in practice.

In order to compare these different methods, we use ROC curves which represent the true positive rate (TPR) as a function of the false positive rate (FPR). Since the variables selected by the sPLSDA method are not assigned to a given modality of the qualitative variable, we will consider that a variable is a true positive as soon as it is selected, giving sPLSDA an advantage.

In Figure 4.1, we observe that in the case of a dependence of the AR(1) type, taking into account the dependence produces better results than methods that consider the columns of the $E$ matrix as independent. Moreover, the higher the level of parsimony, the smaller the differences between the methods and the higher the SNR, the better the performance of the different approaches.

### 4.3.1.2. *Performance according to the choice of model*

In this section, we study the performance of the *stability selection* method described in section 4.2.2.2. Figure 4.2 represents the TPR and FPR for different $N$ values and different thresholds. We observe that when $N$ is greater than 1,000 and when the threshold is equal to 0.95, the FPR is very low and the rate of true positives is very high, regardless of the scenario under study.

The circles (•) in Figure 4.3 show the positions of the variables selected by our method for two threshold values: 0.95 and 1 and $N = 1,000$ replications. The positions of the non-zero coefficients of $B$ are represented by crosses (+). The observations $Y$ are generated with the parameters given at the beginning of section 4.3 for the example of an AR(1) process with $\phi_1 = 0.9$ and $\kappa = 1$. We observe in this figure that the positions of the non-zero coefficients are found much more frequently for the threshold 0.95 than for the threshold 1 and that FPR remain rare even for the 0.95 threshold.

### 4.3.2. *Numerical performance*

In order to study the algorithmic complexity of our method, we generated matrices $Y$ verifying model [4.1] with $n = 30$, a parsimony level of $B$ equal to 0.01 and $q = 100, 1,000, 3,000 \ldots, 5,000$. In addition, we have also generated the rows $E$ as realizations of an AR(1) process. Figure 4.4 represents the execution times of MultiVarSel for different numbers of replications in the *stability selection* step.

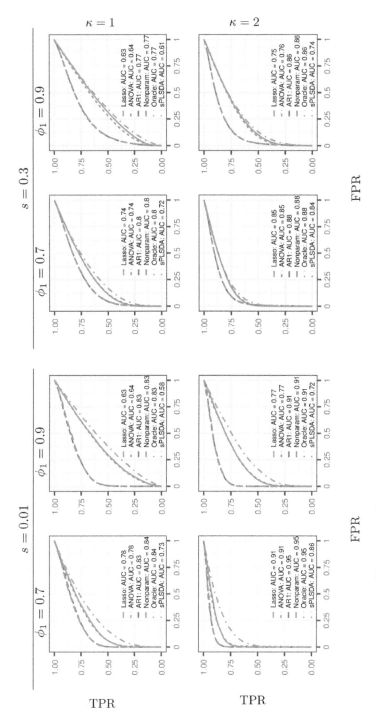

**Figure 4.1.** Empirical means of ROC curves obtained from 200 simulations for the different methodologies for an AR(1) model. On the first line, κ=1, on the second line: κ=2, $\phi_1$ is the AR(1) parameter and s is the degree of parsimony

102  Biological Data Integration

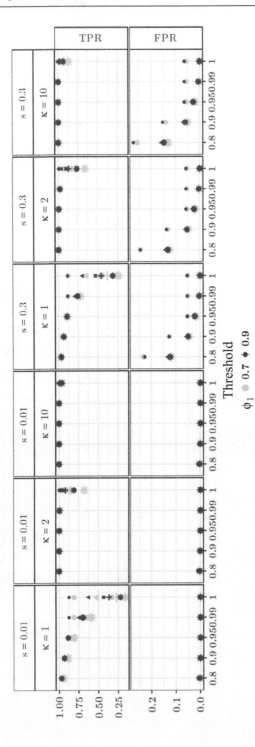

**Figure 4.2.** *Impact of the number $N$ of replications, of the threshold of the stability selection and of the parameter $\phi_1$ of the AR(1) process*

These execution times were obtained for a computer with 16 GB of RAM and 8 Intel Core i7 (3.66GHz) type cores. According to this figure, MultiVarSel takes only a few seconds to analyze a matrix with 5,000 columns.

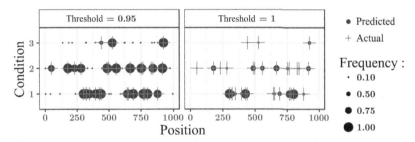

**Figure 4.3.** *Positions of the variables selected by our approach (•) when $\kappa = 1$*

COMMENTS ON FIGURE 4.3.– *The values on the y-axis correspond to the three conditions. The results obtained when the threshold is equal to 0.95 can be found on the left-hand side and those obtained when the threshold is equal to 1 on the right-hand side. A larger dot size implies higher selection frequency.*

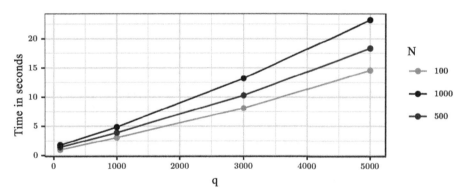

**Figure 4.4.** *Execution time in seconds of MultiVarSel according to the number of columns $q$ of the matrix of observations $Y$. The number of replications corresponds to the number $N$ of subsamples in the stability step selection*

## 4.4. Application to the study of seed quality

In this section, we give the results obtained by applying the R package MultiVarSel to the omics data to understand the influence of the production temperature of Arabidopsis seeds on their protein and metabolite composition. The R commands to be used to study the metabolomic and proteomic data are described in sections 4.6.1 and 4.6.2, respectively.

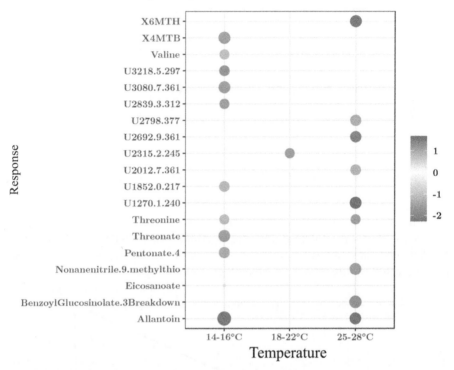

**Figure 4.5.** *Estimation of the coefficients $B_{i,j}$ for the selected metabolites with a threshold equal to 0.93. It should be noted that the area of the circles becomes larger when the absolute values of the coefficients are high*

### 4.4.1. Metabolomics data

We have represented in Figure 4.5 the estimates of the coefficients $B_{i,j}$ of matrix $B$ obtained using our method with a threshold equal to 0.93 and the *boxplots* of the abundances (centered and reduced) of the metabolites selected in Figure 4.6. It should be noted that this threshold was chosen to, on the one hand, limit the number of selected metabolites and, on the other hand, keep the metabolites presenting a biological interest. As such, 19 metabolites were selected, including two glucosinolates, X6MTH (6-methylthiohexyl glucosinolate) and X4MTB (4-methylthiobutyl glucosinolate), which are more abundant in dry mature seeds when the production temperature is high. Conversely, two products of glucosinolate catabolism follow an opposite accumulation profile, since they are characteristic of seeds produced at low temperatures. Glucosinolates are specialized sulfur-rich metabolites containing a glucose molecule and a variable aglycone group. They are involved in the protection of plants against pests and may exhibit antifungal and antioxidant properties (Sønderby et al. 2010). Therefore, the seed production temperature modifies the metabolism of glucosinolates in *Arabidopsis* and consequently their biochemical and physiological quality.

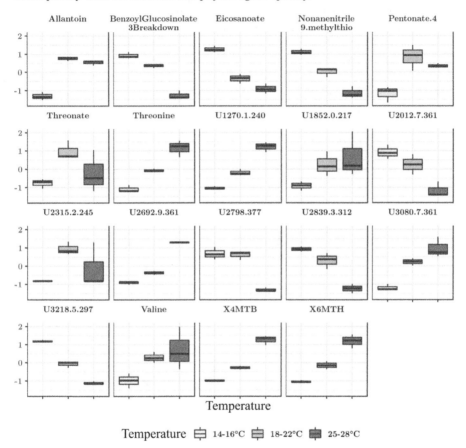

**Figure 4.6.** *Abundance boxplots (centered and reduced) of selected metabolites with a threshold equal to 0.93*

### 4.4.2. Proteomics data

In Figure 4.7, we have represented the estimates of the coefficients $B_{i,j}$ of the matrix $B$ obtained using our method with a threshold equal to 0.95 and *boxplots* of protein abundance selected in Figure 4.8. The results revealed seven proteins characterizing the seeds produced at low temperatures. They are involved in quite different biological and molecular functions. Among them, we can find the protein encoded by the At1g07985 gene. This gene has barely been studied in plants. It codes for a small protein of 16.4 kDa. A sequence analysis via PROSITE (https://prosite.expasy.org/prosite.html) reveals the presence of a bipartite (two-part) nuclear localization signal (NLS_BP, PS50079) for this protein. It has been proposed that the At1g07985 gene is under the control of a bidirectional promoter also involved in the regulation of the At1g07980 gene coding for the transcription factor NF-YC10 (nuclear factor Y, subunit C10) (Kourmpetli et al. 2013). This promoter region contains several G-box elements involved in seed development and response to abscisic acid (ABA) (Keddie et al. 1994). ABA is an important phytohormone for filling, desiccation tolerance and inducing seed dormancy.

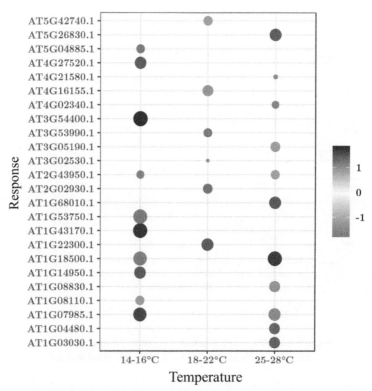

**Figure 4.7.** *Estimation of the coefficients $B_{i,j}$ for the selected proteins with a threshold equal to 0.95. It should be noted that the larger the area of the circles, the higher the absolute value of the coefficients*

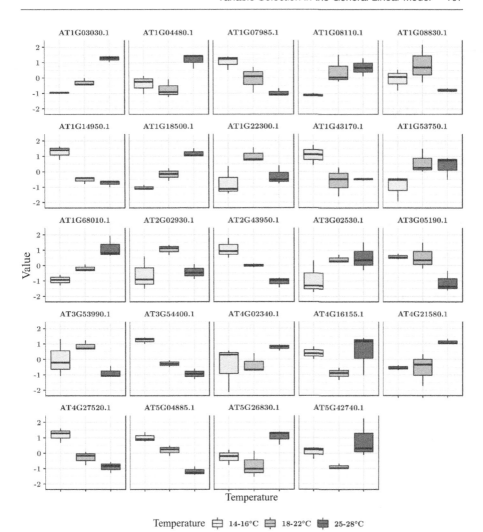

**Figure 4.8.** *Abundance boxplots of selected proteins with a threshold equal to 0.95*

Seeds produced at high temperature are also characterized by seven proteins. Two of them are involved in the translational machinery: the ribosomal protein L14p/L23e (At1g04480) and threonyl-tRNA synthetase (At5g26830). It was demonstrated that the synthesis of new proteins is essential for the germination and development of young plants. Germination specific and essential proteins are derived from the selective and sequential translation of mRNAs stored in the seed and neotranscribed mRNA during imbibition to allow for successful germination and growth of the future seedling. During imbibition, the genes At1g04480 and At5g26830 are upregulated during the germination program stricto sensu. The accumulation of proteins encoded by these genes in mature dry seeds appears to be an indicating element of germination potential.

## 4.5. Conclusion

In this chapter, we propose a new method for variable selection in the general linear model, taking into account the dependence that may exist between the columns of the observation matrix.

Our approach is implemented in the R package MultiVarSel, which is available through the CRAN. We have shown that our method presented a very good statistical and numerical performance, which makes it perfectly well suited for analyzing large-scale omics data.

## 4.6. Appendices

### 4.6.1. *Example of using the package* MultiVarSel *for metabolomic data analysis*

```
require(MultiVarSel) # Loading of package MultiVarSel

data("metabolomAth") # allows loading the metab data table in R

# Definition of matrices X and Y:
temperature <- metab$temperature
Y <- as.matrix(metab[, - 1])
X <- model.matrix(lm(Y ~ temperature + 0, data = metab))
p <- ncol(X)
n <- nrow(X)
q <- dim(Y)[2]
Y <- scale(Y) # Renormalization of matrix Y
```

Here, the matrix $Y$ is renormalized to force the empirical mean of the columns to be zero and the empirical variance of the columns to be equal to 1.

```
residuals <- lm(as.matrix(Y) ~ X - 1)$residuals # Definition of residuals

pvalue <- whitening_test(residus) # Whitening test
print(pvalue)

## [1] 0.03038582
```

This whitening test is used to determine whether dependence is present in the data. The $p$-value is less than 0.05, thereby at the 5% level we have highlighted dependence in the data that needs to be removed.

Several types of dependences can be tested.

```
whitening_choice(residuals, typeDeps = c("AR1", "nonparam"), pAR = 1, qMA = 0)
##            Pvalue       Decision
## AR1         0.035 NO WHITE NOISE
## nonparam    0.741    WHITE NOISE
```

Therefore, the model with the highest $p$-value is chosen. Here, it is the "nonparam" model, that is, the one corresponding to modeling the dependence by a symmetrical Toeplitz matrix.

```
## Computation of the estimator of the square root of the inverse of Sigma
square_root_inv_hat_Sigma <- whitening(residuals, "nonparam", pAR = 1, qMA = 0)

## Computation of the selection frequency of each variable using Lasso
# and "selection frequency"
require(doMC) # to parallelize the computation (does not work on windows)
registerDoMC(cores = 4) # 4 corresponds to the number of cores used

Freqs <- variable_selection(Y, X, square_root_inv_hat_Sigma, nb_repli = 5000,
                            parallel = TRUE, nb.cores = 4)
# 5000 corresponds the number of replications used

threshold <- 0.93 # Threshold choice for the frequencies of the selected variables
indices <- which(Freqs$frequency >= threshold)

## Estimation of selected coefficients in B:
Yvec <- as.numeric(Y %*% square_root_inv_hat_Sigma)
Xvec <- kronecker(t(square_root_inv_hat_Sigma), X)
Xvec_sel <- Xvec[, indices]
B_sel_hat <- solve(t(Xvec_sel) %*% Xvec_sel, t(Xvec_sel) %*% Yvec)
Freqs$estim <- rep(0, p * q)
Freqs$estim[indices] <- as.vector(B_sel_hat)

## Plot of selected metabolites with the estimation of their
# associated B coefficient
ggplot(data = Freqs[Freqs$frequency >= threshold, ],
       aes(y = Names_of_Y, x = Names_of_X,
           color = estim, size = abs(estim))) +
  geom_point() +
  scale_size_continuous(name = "", breaks = c(0, 0.5, 1, 2, 3),
                        guide = FALSE)+
  scale_color_gradient2(low = "steelblue", mid = "white", high = "darkred") +
  labs(y = "Response", x = "Temperature", color = "")
```

This graph is represented in Figure 4.5.

We now give the commands to obtain boxplots similar to those obtained in Figure 4.6.

```
require(reshape2) # package used to transform the data

## Boxplots of selected metabolites
# The database is adapted for the graph
metab_sel <- as.character(Freqs[Freqs$frequency >= threshold, "Names_of_Y"])
temperature <- metab$temperature
metab_scale <- cbind.data.frame(temperature, Y[, metab_sel])
mm_sel <- melt(metab_scale, id.vars = "temperature")
# Graph
ggplot(data = mm_sel, aes(x = temperature, y = value, fill = temperature)) +
  geom_boxplot() +
  facet_wrap(~variable) +
  theme(axis.text.x = element_blank()) +
  labs(y = "value", x = "Temperature", fill = "Temperature")+
  scale_fill_brewer(palette="Blues")
```

### 4.6.2. Example of using the package `MultiVarSel` for proteomic data analysis

```
require(MultiVarSel) # Loading of package MultiVarSel

data("proteomAth") # allows loading the prot data table into R

# Definition of matrices X and Y:
temperature <-prot$temperature
Y <- as.matrix(prot[, - 1])
X <- model.matrix(lm(Y ~ temperature + 0, data = prot))
p <- ncol(X)
n <- nrow(X)
q <- dim(Y)[2]
Y <- scale(Y) # Renormalization of matrix Y
```

Here, the matrix $Y$ is renormalized to force the empirical mean of the columns to be zero and the empirical variance of the columns to be equal to 1.

```
residuals <- lm(as.matrix(Y) ~ X - 1)$residuals # Definition of residuals

pvalue <- whitening_test(residus) # Whitening test
print(pvalue)

## [1] 0.06240354
```

This whitening test results in determining whether dependence is present in the data. The *p*-value is less than 0.05, thereby at the 5% level we have highlighted dependence in the data that needs to be removed.

Several types of dependences can be tested: here AR(1) and symmetric Toeplitz (the most general case noted here "nonparam").

```
whitening_choice(residuals, typeDeps = c("AR1", "nonparam"), pAR = 1, qMA = 0)

##             Pvalue    Decision
## AR1          0.187    WHITE NOISE
## nonparam         1    WHITE NOISE
```

Therefore, the model with the highest $p$-value is chosen.

```
## Computation of the estimator of the square root of the inverse of Sigma
square_root_inv_hat_Sigma <- whitening(residuals, "nonparam", pAR = 1, qMA = 0)

## Computation of the selection frequency of each variable using Lasso
# and "stability selection"
require(doMC)   # to parallelize the computation (does not work on windows)
registerDoMC(cores = 4)  # 4 corresponds to the number of cores used


Freqs <- variable_selection(Y, X, square_root_inv_hat_Sigma, nb_repli = 5000,
                    parallel = TRUE, nb.cores = 4)
# 5000 corresponds the number of replications used
threshold <- 0.95 # Threshold choice for the frequencies of the selected variables
indices <- which(Freqs$frequency >= threshold)

## Estimation of selected coefficients in B:
Yvec <- as.numeric(Y %*% square_root_inv_hat_Sigma)
Xvec <- kronecker(t(square_root_inv_hat_Sigma), X)
Xvec_sel <- Xvec[, indices]
B_sel_hat <- solve(t(Xvec_sel) %*% Xvec_sel, t(Xvec_sel) %*% Yvec)
Freqs$estim <- rep(0, p * q)
Freqs$estim[indices] <- as.vector(B_sel_hat)

## Plot of the selected proteins with the estimation of their
# associated B coefficient
ggplot(data =  Freqs[Freqs$frequency >= threshold, ],
            aes(y = Names_of_Y, x = Names_of_X,
                color = estim, size = abs(estim))) +
  geom_point() +
  scale_size_continuous(name = "", breaks = c(0, 0.5, 1, 2, 3),
                    guide = FALSE)+
  scale_color_gradient2(low = "steelblue", mid = "white", high = "darkred") +
  labs(y = "Response", x = "Temperature", color = "")
```

This graph is represented in Figure 4.7.

We now give the commands to obtain boxplots similar to those obtained in Figure 4.8.

```
require(reshape2) # package used to transform the data

## Boxplots of selected proteins
# The database is adapted for the graph
prot_sel <- as.character(Freqs[Freqs$frequency >= threshold, "Names_of_Y"])
temperature <- prot$temperature
prot_scale <- cbind.data.frame(temperature, Y[, prot_sel])
mp_sel <- melt(prot_scale, id.vars = "temperature")
# Graph
ggplot(data = mp_sel, aes(x = temperature, y = value, fill = temperature)) +
  geom_boxplot() +
  facet_wrap(~variable) +
  theme(axis.text.x = element_blank()) +
  labs(y = "value", x = "Temperature", fill = "Temperature")+
  scale_fill_brewer(palette="Blues")
```

We propose here another way to create Figures 4.7 and 4.8 using the R package tidyverse.

```
require(tidyverse)
Yvec <- as.numeric(Y %*% square_root_inv_hat_Sigma)
Xvec <- kronecker(t(square_root_inv_hat_Sigma), X)
colnames(Xvec) <-  paste(rep(colnames(Y), each = ncol(X)),
                    rep(colnames(X), ncol(Y)), sep = "_")

# recovery of selected protein-temperature associations:
sel  <- Freqs[Freqs$frequency >= threshold, c("Names_of_Y","Names_of_X")] %>%
  unite("sel") %>% pull(sel)

# recovery of the selected proteins:
sel_prot <- Freqs[Freqs$frequency >= seuil, "Names_of_Y"] %>%
  as.character()

# Estimate of B
Xvec_sel <- as.matrix(Xvec)[,sel]
B_sel_hat <- solve(t(Xvec_sel) %*% Xvec_sel, t(Xvec_sel) %*% Yvec)

# Graph representing the non-zero values of B
p <- B_sel_hat %>% as.data.frame() %>%
  rownames_to_column() %>%
  mutate(rowname = str_remove_all(rowname,"sel")) %>%
  separate(rowname, into = c("Response", "Temperature" ), sep="_") %>%
  mutate(Temperature= str_remove( Temperature,"temperature")) %>%
  ggplot(aes(y = Response, x = Temperature, color = V1, size = abs(V1))) +
  geom_point() +
  scale_size_continuous(name = "", breaks = c(0, 0.5, 1, 2, 3))+
  scale_color_gradient2(low = "steelblue",mid = "white", high = "darkred") +
  labs(y = "Response", x = "Temperature", color = "")
```

```
# Graph representing the boxplots of the selected proteins
p <- Y %>% as.data.frame() %>%
  select(sel_prot) %>%
  mutate(temperature = temperature) %>%
  gather(key = "Response", value = value, - temperature) %>%
  ggplot(aes(x = temperature, y = value, fill = temperature)) +
  geom_boxplot() +
  facet_wrap( ~ Response) +
  theme(axis.text.x = element_blank(), legend.position = "bottom") +
  labs(y = "valeur", x = "Temperature", fill = "Temperature") +
  scale_fill_brewer(palette = "Blues")
```

## 4.7. Acknowledgments

We would to thank the whole *consortium* of the European project EcoSeed (*FP7 Environment, Grant/Award Number: 311840 EcoSeed*). IJPB is supported by the LABEX Saclay Plant Sciences-SPS (ANR-10-LABX-0040-SPS). We would like to thank all the people who contributed to the production of the biological material and proteomic and metabolomic data. More particularly, we would like to thank the University of Warwick (UWAR, Finch-Savage, W.E. and Awan, S.) for the production of seeds, the biochemistry platform of the IJPB Plant Observatory (OV-Biochemistry, Bailly, M.) for the management and preparation of the samples for proteomics and metabolomics, the proteomics analysis platform of Paris Sud-Ouest (PAPPSO, Balliau, T. and Zivy, M.) for proteome mass spectrometry analyses and the chemistry-metabolism platform of the Observatoire du végétal of the IJPB (OV chemistry-metabolism, Clément, G.) for GC/MS metabolome analysis.

## 4.8. References

Bailey-Serres, J., Sorenson, R., Juntawong, P. (2009). Getting the message across: Cytoplasmic ribonucleoprotein complexes. *Trends in Plant Science*, 14(8), 443–453.

Blödner, C., Goebel, C., Feussner, I., Gatz, C., Polle, A. (2007). Warm and cold parental reproductive environments affect seed properties, fitness, and cold responsiveness in arabidopsis thaliana progenies. *Plant, Cell & Environment*, 30(2), 165–175.

Boccard, J. and Rudaz, S. (2016). Exploring omics data from designed experiments using analysis of variance multiblock orthogonal partial least squares. *Analytica Chimica Acta*, 920, 18–28.

el Bouhaddani, S., Uh, H.-W., Hayward, C., Jongbloed, G., Houwing-Duistermaat, J. (2018). Probabilistic partial least squares model: Identifiability, estimation and application. *Journal of Multivariate Analysis*, 167, 331–346.

Brockwell, P. and Davis, R. (1991). *Time Series: Theory and Methods*. Springer-Verlag, New York.

Burghardt, L.T., Edwards, B.R., Donohue, K. (2016). Multiple paths to similar germination behavior in arabidopsis thaliana. *New Phytologist*, 209(3), 1301–1312.

Durif, G., Modolo, L., Michaelsson, J., Mold, J.E., Lambert-Lacroix, S., Picard, F. (2017). High dimensional classification with combined adaptive sparse PLS and logistic regression. *Bioinformatics*, 34(3), 485–493.

Faraway, J.J. (2004). *Linear Models with R*. Chapman & Hall/CRC, Boca Raton, FL.

Galland, M., Huguet, R., Arc, E., Cueff, G., Job, D., Rajjou, L. (2014). Dynamic proteomics emphasizes the importance of selective mRNA translation and protein turnover during arabidopsis seed germination. *Molecular & Cellular Proteomics*, 13(1), 252–268.

Keddie, J.S., Tsiantis, M., Piffanelli, P., Cella, R., Hatzopoulos, P., Murphy, D.J. (1994). A seed-specific brassica napus oleosin promoter interacts with a G-box-specific protein and may be bi-directional. *Plant Molecular Biology*, 24(2), 327–340.

Kourmpetli, S., Lee, K., Hemsley, R., Rossignol, P., Papageorgiou, T., Drea, S. (2013). Bidirectional promoters in seed development and related hormone/stress responses. *BMC Plant Biology*, 13(1), 187.

Lê Cao, K.-A., Boitard, S., Besse, P. (2011). Sparse PLS discriminant analysis: Biologically relevant feature selection and graphical displays for multiclass problems. *BMC Bioinformatics*, 12(1), 253.

Mardia, K., Kent, J., Bibby, J. (1979). *Multivariate Analysis*. Academic Press, London.

Meinshausen, N. and Buhlmann, P. (2010). Stability selection. *Journal of the Royal Statistical Society*, 72(4), 417–473.

Muller, K.E. and Stewart, P.W. (2006). *Linear Model Theory: Univariate, Multivariate, and Mixed Models*. John Wiley & Sons, Hoboken, NJ.

Perrot-Dockès, M., Lévy-Leduc, C., Sansonnet, L., Chiquet, J. (2018). Variable selection in multivariate linear models with high-dimensional covariance matrix estimation. *Journal of Multivariate Analysis*, 166, 78–97.

Perrot-Dockès, M., Lévy-Leduc, C., Rajjou, L. (2019). Estimation of large block structured covariance matrices: Application to "multi-omic" approaches to study seed quality. arXiv:1806.10093v2.

Ren, S., Hinzman, A.A., Kang, E.L., Szczesniak, R.D., Lu, L.J. (2015). Computational and statistical analysis of metabolomics data. *Metabolomics*, 11(6), 1492–1513.

Rothman, A.J., Levina, E., Zhu, J. (2010). Sparse multivariate regression with covariance estimation. *Journal of Computational and Graphical Statistics*, 19(4), 947–962.

Saccenti, E., Hoefsloot, H.C.J., Smilde, A.K., Westerhuis, J.A., Hendriks, M.M.W.B. (2013). Reflections on univariate and multivariate analysis of metabolomics data. *Metabolomics*, 10(3), 361–374.

Sønderby, I.E., Geu-Flores, F., Halkier, B.A. (2010). Biosynthesis of glucosinolates – Gene discovery and beyond. *Trends in Plant Science*, 15(5), 283–290.

Springthorpe, V. and Penfield, S. (2015). Flowering time and seed dormancy control use external coincidence to generate life history strategy. *Elife*, 4, e05557.

Tibshirani, R. (1996). Regression shrinkage and selection via the Lasso. *J. Royal. Statist. Soc B.*, 58(1), 267–288.

Zhang, H., Zheng, Y., Yoon, G., Zhang, Z., Gao, T., Joyce, B., Zhang, W., Schwartz, J., Vokonas, P., Colicino, E. et al. (2017). Regularized estimation in sparse high-dimensional multivariate regression, with application to a DNA methylation study. *Stat. Appl. Genet. Mol. Biol.*, 16(3), 159–171.

# 5

# Structured Compression of Genetic Information and Genome-Wide Association Study by Additive Models

Florent GUINOT[1,2], Marie SZAFRANSKI[2,3]
and Christophe AMBROISE[2]

[1] *Institut Roche, Boulogne-Billancourt, France*
[2] *LaMME, Université Paris-Saclay, CNRS, Université d'Évry,*
*Évry-Courcouronnes, France*
[3] *ENSIIE, France*

In this chapter, we present an extension of the approach proposed by Guinot et al. (2018). The LEOS (LEarning the Optimal Scale in GWAS) method is available at: http://stat.genopole.cnrs.fr/leos.

In the first instance, in section 5.1, we will look into the context of genome-wide association studies (GWAS) and the reference tools used to detect associations between a genotype and a phenotype. In section 5.2, we will then describe our method that exploits the structure of linkage disequilibrium (LD) in genomes to

---

For a color version of all figures in this chapter, see: http://www.iste.co.uk/froidevaux/biologicaldata.zip.

improve the statistical power in genome-wide association studies. Finally, in section 5.3, we will illustrate the results obtained in a study on ankylosing spondylitis (AS).

## 5.1. Genome-wide association studies

### 5.1.1. *Introduction to genetic mapping and linkage analysis*

**Genetic mapping** is based on the use of genetic markers for constructing maps showing the position of genes and other sequences on a genome. Historically, the first markers used to construct genetic maps were genes encoding Mendelian traits (highly heritable qualitative traits), with distinct phenotypes for each allele (see Sturtevant (2001) for more details on early genetic mapping work). However, although genes are relevant markers for mapping small genomes, genetic maps based on these alone lack accuracy when mapping larger genomes, this being partly due to the large non-coding regions that can separate two coding regions.

Genetic mapping makes use of the principle of heredity described by Mendel (1865) and the resulting properties of genetic linkage to estimate the relative position of each DNA marker on a chromosome. The principle of genetic linkage arises from the fact that, since chromosomes are inherited as intact units, alleles of genes located on the same chromosome should also be inherited together. This principle, derived from Mendel's second law and stipulating that pairs of alleles should separate independently, is not what is observed in reality. In fact, genetically related genes are sometimes inherited together and sometimes not, resulting in a phenomenon of partial genetic linkage.

This partial binding property is explained by the behavior of chromosomes during meiosis, during which homologous chromosomes can undergo physical breakage and exchange DNA fragments within the context of recombination (*crossing-over*). These recombination events explain why linked genes, and thus linked DNA markers, are not always transmitted together (thus disregarding Mendel's second law). It is this phenomenon that makes it possible to map the relative position of markers on a genome, since two markers that are genetically close will have a lower probability of being separated during meiosis than two markers that are far apart. Consequently, the probability that two markers on the same chromosome will recombine is a function of the distance between them. This recombination frequency is therefore an indirect measure of the distance between two markers and it is thus possible to map the relative positions of each marker by estimating these recombination frequencies within a population.

Comparisons between genetic maps and the actual positions of genes on the genome have shown that certain chromosome regions, called **recombination**

**hotspots**, are more likely to be involved in *cross-over* than others. Despite the fact that every individual has a unique sequence, this results in chromosomal regions shared by the same population known as **haplotypic genome structure**.

**Linkage analysis** is the traditional approach for mapping genes involved in diseases where the cosegregation of disease-associated alleles is studied within larger or smaller families. Although this approach proves effective in locating genes contributing to simple Mendelian disorders, it is unreliable when it comes to mapping complex diseases. Actually, for these diseases several genes may be interacting and the effects of the genes involved may vary according to environmental and other non-genetic risk factors.

### 5.1.2. Principles of genome-wide association studies

Genome-wide association studies focus on identifying genetic markers that appear in a representative sample of unrelated individuals at different frequencies between individuals affected by a disease and healthy individuals chosen as controls.

These studies use the fact that it is easier to establish large cohorts of people sharing a genetic risk factor for a complex disease across the whole population rather than within families, as is the case in traditional association studies. Genome-wide association studies rely on two types of association: **direct** and **indirect** association.

On the one hand, direct association focuses on genotyping and studying functional polymorphisms with a high probability of disrupting a biological system, such as non-synonymous mutation[1], splice site variants[2] or the copy number of a polymorphism (CNP[3]).

On the other hand, indirect association looks at functional polymorphisms, such as non-synonymous mutations, but also polymorphisms that are close to them. As a matter of fact, even if the latter are unlikely to be directly associated with the phenotype, at a sufficiently high density, one or more may be highly correlated (that is, in LD) with the underlying causative mutations.

---

1. A non-synonymous mutation modifies the protein sequence as opposed to a synonymous mutation.
2. Genetic modification of the DNA sequence that occurs at the boundary of an exon and intron (splice site). They can disrupt RNA splicing, causing the loss of exons or inclusion of introns, leading to a change in protein coding.
3. A CNP is a normal variation in DNA due to the varying number of copies of a sequence within the DNA. CNP variants are common and widely distributed in the genome.

In addition, due to recent advances in microarray technology, it is now possible to characterize an individual's genome with several hundred thousand genetic markers (*single nucleotide polymorphism* (SNP)). Within this context, genome-wide association studies have been widely used to identify causative genomic variants[4] involved in the expression of different human diseases (rare, Mendelian or multifactorial diseases). It is now possible to genotype the complete DNA sequence of an individual at a moderate cost, around $1,000 in 2016 (Wetterstrand 2016), and in a very short time. It is thus reasonable to think that SNPs will be abandoned in favor of complete genotyping, and it is therefore necessary to develop statistical methods capable of processing this type of massive data. We will discuss the classical statistical methods used in association studies in section 5.1.5.

### 5.1.3. *Single nucleotide polymorphism*

An SNP occurs at a specific location in the genome (see Figure 5.1). In a given population, most individuals have a specific nucleotide at one position (a cytosine, for example) but a minority of individuals may exhibit a different nucleotide at that same position (a thymine, for example). The two possible nucleotide variations at a given position in the genome (*locus*) are said to be **alleles**; this type of polymorphism is extremely frequent in the human genome (a few million). The vast majority of SNPs are biallelic because they originate from a point mutation in the genome, converting one nucleotide into another. For an SNP to be more than biallelic, it would be necessary for a new mutation appear, after the first one has been fixed in the population, at exactly the same position in the genome, which is highly unlikely. SNP genotyping methods are based on oligonucleotide hybridization analysis, where an oligonucleotide (short single-stranded DNA molecule used as a probe) will only be hybridized with another DNA molecule (called a **target**) if the two molecules are able to form a complementary structure, under specific temperature conditions.

Oligonucleotide hybridization makes it possible to distinguish between the two alleles of a SNP if there is at least one difference at a given position between the probe oligonucleotide and the target DNA. There are several existing screening methods based on oligonucleotide hybridization: namely, a DNA chip which uses fluorescent markers to detect hybridization, the oligonucleotide ligation assay (OLA) by capillary electrophoresis and the amplification refractory mutation system (ARMS) based on PCR primers and electrophoresis.

---

4. We will now equally refer to the terms "variant", "marker", "SNP" or "polymorphism" to designate the variable of interest in association studies.

Recent advances in microarray technology have allowed the genotyping of SNPs at a moderate cost. As a result, it has become possible to characterize the genome of a individual by several million genetic markers. DNA chip-based technology uses a piece of glass or silicon (see Figure 5.2). In order to prepare very high density chips, the oligonucleotides are synthesized in situ on the surface of the glass. A density of up to 700,000 oligonucleotides per cm$^2$ is possible and 150,000 polymorphism marks can be typed into a single experiment (Brown 2007).

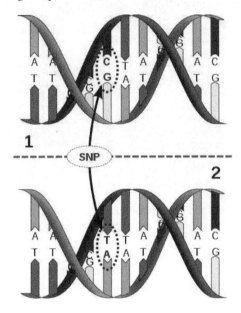

**Figure 5.1.** *Schematic representation of nucleotide polymorphism (© David Hall/Creative Commons Licence)*

DNA chips are presently used to diagnose a number of diseases. Despite being easy to implement and incurring lower costs, microarrays are gradually being replaced by new DNA sequencing technologies whose applications can be applied in the field of diagnostics. Nevertheless, the use of microarrays is still preferred in very large studies, as well as in some clinical tests. For example, the first-ever FDA-approved diagnostic microarray – the Amplichip CYP450 based on Affymetrix technology – was designed to identify the different alleles of two genes in the cytochrome P450 complex, the CYP2D6 and CYP2C19 genes, which are involved in the metabolism of many psychotropic drugs (Chneiweis 2007).

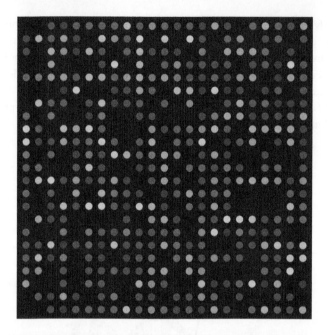

**Figure 5.2.** *Visualization of DNA microarray hybridization by fluorescence (Guillaume Paumier, licensed under CC BY-SA 3.0)*

COMMENTS ON FIGURE 5.2.– *The DNA to be tested (or target) is marked with a fluorochrome and put in contact with the surface of the microchip. Hybridization is detected with a fluorescence microscope, and the position of the fluorescent signal indicates which oligonucleotide probes have hybridized with the target DNA (Brown 2007).*

### 5.1.4. *Disease penetrance and* odds ratio

Considering a biallelic locus with alleles $A$ and $a$, the possible genotypes are $A/A$, $A/a$ and $a/a$. The **disease penetrance** associated with a given genotype is the risk of disease in individuals carrying this genotype. Assuming a genetic penetrance of parameter $\gamma > 1$, the main models of disease penetrance in the genetics of association are as follows:

– the **multiplicative model**: the risk of disease is increased by a factor $\gamma$ with each additional allele $a$;

– the **additive model**: the risk of disease is increased by a factor $\gamma$ for the genotype $A/a$ and by a factor of $2\gamma$ for the genotype $a/a$;

– the **recessive model**: the risk of disease is increased by a factor $\gamma$ for the genotype $a/a$ only;

– the **dominant model**: the risk of disease is increased by a factor $\gamma$ for both the genotypes $A/a$ and $a/a$.

With a disease penetrance model, the genotype of an individual for a particular SNP can be represented by a numerical value as a function of the number of minor alleles, that is of lower frequency in the population it carries. Table 5.1 gives SNP coding in different penetrance models.

| Genotype | Additive component $z_{(A)i}$ | Dominant component $z_{(D)i}$ | Recessive component $z_{(R)i}$ |
|---|---|---|---|
| $A/A$ | 0 | 0 | 0 |
| $A/a$ | 1 | 1 | 0 |
| $a/a$ | 2 | 1 | 1 |

**Table 5.1.** *Additive, dominant and recessive SNP coding*

A commonly used indicator for measuring the strength of an association between phenotype and genotype is **relative risk** (RR), which compares disease penetrance between individuals carrying different genotypes. To estimate RR, it is therefore necessary to assess disease penetrance. This can be derived directly from prospective cohort studies. In these studies, a group of exposed individuals (cases) and a group of unexposed individuals (controls) from the same population are followed and compared. Nonetheless, for case–control studies in which the ratio of cases to controls is determined by the investigator, it is impossible to achieve estimates of disease penetrance and therefore of RRs.

| | Genotypic counts | | | | Allelic counts | | |
|---|---|---|---|---|---|---|---|
| | $A/A$ | $A/a$ | $a/a$ | Total | $A$ | $a$ | Total |
| Case | $n_{01}$ | $n_{11}$ | $n_{21}$ | $n_{.1}$ | $m_{01}$ | $m_{11}$ | $m_{.1}$ |
| Controls | $n_{00}$ | $n_{10}$ | $n_{20}$ | $n_{.0}$ | $m_{00}$ | $m_{10}$ | $m_{.0}$ |
| Total | $n_{0.}$ | $n_{1.}$ | $n_{2.}$ | $N$ | $m_{0.}$ | $m_{1.}$ | $N$ |

**Table 5.2.** *Genotypic and allelic contingency tables for a locus characterized by two possible $A$ and $a$ alleles*

COMMENTS ON TABLE 5.2.– *Genotypic counts $n_{ij}$ correspond to the number of individuals carrying $i$ copies of the minor allele $a$ and phenotype $j = 1$ for cases and $j = 0$ for controls. The allelic counts $m_{ij}$ can be summarized in different ways according to the disease penetrance model: for the dominant model, $i = 0$ if a person* is A/A and $i = 1$, whereas for a recessive model $i = 0$ if a person is A/A or A/a and $i = 1$.

In this type of study, the strength of an association is measured by the odds ratio (OR) (Clarke et al. 2011). In these studies, two types of ORs can be calculated:

– the allelic OR is estimated by comparing the disease probability for an individual carrying the $A$ allele to that of an individual carrying the $a$ allele;

– the genotypic OR is estimated by comparing the disease probability for an individual carrying a certain genotype to that of an individual carrying another genotype.

### 5.1.5. *Single marker analysis*

The classic approach to identifying variants associated with disease tests the effect of each SNP one by one, using a strategy based on hypothesis testing. The principle consists of identifying variants statistically associated with the phenotype, these variants being in LD themselves, with potential causal polymorphism. This section presents the most commonly used tests.

#### 5.1.5.1. *Pearson's $\chi^2$ statistic*

Under the assumption of independence between phenotype and marker, the expectation of the genotypic count $n_{ij}$ (see Table 5.2) is denoted by:

$$\mathbb{E}(n_{ij}) = \frac{n_i.n_{.j}}{N}$$

It is thus possible to perform a genotypic association test by testing the row-column independence of the contingency table by means of a conventional $\chi^2$ test, whose decision statistic is:

$$\chi^2_{genotypic} = \sum_{i=0}^{2}\sum_{j=0}^{1} \frac{(n_{ij} - \mathbb{E}(n_{ij}))^2}{\mathbb{E}(n_{ij})}$$

This test statistic follows a 2 degree-of-freedom (d.o.f.) $\chi^2$ distribution under the null hypothesis of independence.

It is also possible to consider alternative penetrance models based on counts, namely, allelic counts rather than genotypic ones. In this case, the allelic association test is carried out with a contingency table $2 \times 2$, and the test statistic is defined as:

$$\chi^2_{allelic} = \sum_{i=[0,1]}\sum_{j=[0,1]} \frac{(m_{ij} - \mathbb{E}(m_{ij}))^2}{\mathbb{E}(m_{ij})}$$

This allelic association test has a single degree of freedom and will be more powerful than the genotypic test as long as the heterozygous genotype penetrance is intermediate compared to those of the two homozygous genotypes (Clarke et al. 2011).

### 5.1.5.2. Cochran-Armitage trend test

Any penetrance model specifying a trend in risk through an increase of the number of $a$ alleles can be examined using the Cochran–Armitage trend test (Cochran 1954; Armitage 1955), in which the test statistic is defined as being:

$$\chi^2_{CA} = \frac{\left[\sum_{i=0}^{2} w_i(n_{.1}n_{2.} - n_{.2}n_{1.})\right]^2}{\frac{n_{1.}n_{2.}}{n}\left[\sum_{i=0}^{2} w_i^2 n_{.i}(n - n_{.i}) - 2\sum_{j=0}^{1}\sum_{i=j+1}^{2} w_j w_i n_{.j} n_{.i}\right]} \quad [5.1]$$

where $w = (w_0, w_1, w_2)$ are weights chosen to detect types of association. For example, for a dominant model $w = (0, 1, 1)$ is optimal, while for a recessive model $w = (0, 0, 1)$ will be chosen instead.

Under the null hypothesis of no association between SNP and disease ($H_0$ : independence between the contingency table rows and columns), $\chi^2_{CA}$ approximately follows a $\chi^2$ with 1 d.o.f. The power of this test is often improved as long as the risks of disease associated with the $A/a$ genotype are intermediate to those associated with the $a/a$ and $A/A$ genotypes. In a GWAS study where the underlying genetic model is unknown, the additive version of this test, that is, with $w = (0, 1, 2)$, is the most commonly used.

### 5.1.5.3. Limitations

The classical approach of single marker analysis is subject to false positives, that is, the discovery of SNPs falsely identified as being associated, due to the number of tests performed at the same time. In order to overcome this issue, a multiple test correction is conventionally applied. Unfortunately, this correction increases the risk of missing true associations that only have a small impact on the phenotype, which is usually the case in GWAS. In fact, to simultaneously test $10^5$ SNPs using the single-marker approach, the significance threshold of the $p$-value is $5.10^{-5}$ by Bonferroni correction, or a slightly higher threshold with a *false discovery rate (FDR)* control method.

Another commonly used approach to multiple testing in GWAS is based on the concept of genome-wide significance. It is based on the distribution of the LD in the genome for a given population and considers that there are an effective number of independent genomic regions, and thus an effective number of statistical tests. For a population of European origin, the threshold on the $p$-value corresponding to the effective number of tests has been estimated at $7.2 \times 10^{-8}$ (Dudbridge and Gusnanto 2008). However, this approach should be employed with caution since the only scenario where this correction is suitable is for a genome-wide test. Gene-candidate or replication studies involving a very limited number of hypotheses do not require any correction at this level, since the number of efficient and independent statistical tests is much lower than what is assumed for genome-wide significance (Bush and Moore 2012).

Furthermore, as indicated by Maher (2008), these approaches need to address other limitations:

1) Correlations between predictors are not taken into account, whereas these correlations can be very strong as a result of LD. Some SNPs can be correlated, even when they are not physically close, because of population structure or epistasis (gene-gene interactions).

2) Epistasis is not taken into account, namely, the causal effects observed only when certain combinations of simultaneous mutations are present in the genome.

3) It is not possible to estimate the risk of genetic disease through predictive models.

4) Only common markers with minor allele frequencies (MAFs) greater than 5% are taken into consideration, although it likely appears that low frequency ($0.5\% < \text{MAF} < 5\%$) or rare variants ($\text{MAF} < 0.5\%$) may explain additional disease risks or variability in certain traits (Lee et al. 2014).

The discovery of part of the missing heritability can sometimes be achieved by taking correlations between variables, epistasis, into account – but this is rarely feasible within the context of GWAS due to the number of tests (and the necessary correction of multiple tests) – as well as the enormous computational costs of such analyses (Manolio et al. 2009).

### 5.1.6. *Multi-marker analysis*

Within the context of multifactorial disease genetic mapping, we have seen that it was necessary that the genotyped SNPs had to be minimally correlated with the causal polymorphism, as each SNP has a weak individual effect on the phenotype. In this section, we will show that it is possible to use the joint effect of several SNPs close to causal polymorphism to improve the detection potential for performing analysis of rare variants, or when the individual effects are too small to be detected.

SNPs can be grouped in several possible ways in multilocus analysis. We can consider grouping together SNPs related to the same biological context such as a biochemical pathway, a protein family or a gene. Other alternatives consists of working at the haplotype level[5] rather than at the genotype level, or using the genome haplotype structure to define clusters of relevant variants.

Performing multi-locus analysis is not as simple as single-marker analysis and can present many challenges in terms of computations and statistical modeling. The most common model used for modeling the joint effect of multiple SNPs is multiple linear regression, where all SNPs belonging to the same previously defined group can be modeled together. In order to reduce collinearity and overfitting problems that may occur, it is possible to resort to penalized regression-based approaches such as *ridge* or *Lasso* (least absolute shrinkage and selection operator) regression. In addition, it is also possible to examine statistical interactions between genetic variants in order to study epistatic effects (Stanislas et al. 2017).

Other methods using multiple linear regression take into account LD within genes to improve statistical power (Yoo et al. 2016) or rather group low-effect variants around known *loci* in order to increase the percentage of variance explained by genetic effects for complex traits (Pare et al. 2015). Finally, other approaches will summarize test statistics resulting from univariate analyses performed within the same region, for example, the $p$-value combination procedures described by Petersen et al. (2013).

### 5.1.6.1. *Haplotype-based approaches*

One approach used in multilocus analysis consists of focusing on the effects of haplotypes rather than those of genotypes. The human genome can be divided into haplotype blocks where most of the intra-block variability is due to mutation rather than recombination. As a result, much of the common genetic variation can also be structured into haplotypes within blocks of LD that are rarely disrupted by meiosis.

It is common to assume that each of the pairs of haplotypes, forming the diplotype $H_i$ of the $i$th individual, independently contributes to the risk of disease. Under this hypothesis, the contribution of each diplotype haplotype can be coded by its effect relative to the most frequent haplotype. The logistic regression model is then parameterized according to the logarithm of the OR for each haplotype (Balding et al. 2008). The linear predictor $\eta_i$ of individual $i$ can be defined as follows:

$$\eta_i = \beta_0 + \sum_{j=1}^{p} \alpha_j z_{ij} + \sum_{k=2}^{h} \beta_k x_{ik}$$

where $z_{ij}$ is the value of the individual $i$ at the covariate $j$ (age, sex, clinical variables), $\alpha_j$ the corresponding coefficient, $\beta_k$ is the log-OR of the $k$th most frequent haplotype

---

5. A haplotype is a group of alleles from different *loci* located on the same chromosome and usually transmitted together.

with respect to the reference haplotype ($k = 1$) and $x_{ik}$ indicates the number of haplotypes of type $k$ in individual $i$ (0, 1 or 2).

A major issue with this approach is that we do not directly observe the diplotype $H$ from the unphased genotype data. One solution consists of performing an estimate of the diplotype for each individual, using a statistical methodology such as PHASE (Stephens et al. 2001) or by maximum likelihood using the EM algorithm (Excoffier and Slatkin 1995). However, due to the uncertainty in the haplotype reconstruction process, the variances of the model parameters are underestimated, which may lead to inflation of the type-I error (Balding et al. 2008).

### 5.1.6.2. Rare variant analysis

In the context of rare variant association analysis, many region- or gene-based multi-marker tests have been proposed, such as burden tests (Asimit et al. 2012), variance-component tests (Wu et al. 2011) or still tests combining these two approaches (Lee et al. 2012). Instead of testing each marker individually, these methods evaluate the cumulative effects of several SNPs within a region, which results in increasing the statistical power when several markers in the cluster are associated with a given trait.

We first introduce the most commonly used model for the study of rare variants. We assume that individuals have been genotyped for a region comprising $M$ genetic markers and define the linear predictor $\eta_i$ of individual $i$ as follows:

$$\eta_i = \beta_0 + \sum_{j=1}^{p} \alpha_j z_{ik} + \sum_{m=1}^{M} \beta_m x_{im} \quad [5.2]$$

where $x_{im} = x_{(A)im}$ is the variable representing the additive component of individual $i$ for marker $m$ and $z_{ij}$ is the value of individual $i$ for covariate $j$. We define the score statistic for marker $m$ as:

$$S_m = \sum_{i=1}^{n} x_{im}(y_i - \eta_i) \quad [5.3]$$

where $y_i$ is the phenotype of individual $i$.

It should be noted that $S_m$ is positive when the marker $m$ is associated with an increase in disease risk (or in the value of the phenotype in general) and is negative in the opposite case.

### 5.1.6.2.1. Burden tests

*Burden tests* (Li and Leal 2008; Asimit et al. 2012) compute a single genetic score based on several genetic markers and test the association between this score and a phenotype of interest. A simple approach consists of summarizing the genotypic information by counting the number of minor alleles among all the markers comprised in the cluster. The summarized genetic score is then defined as:

$$C_i = \sum_{m=1}^{M} \omega_m x_{im} \qquad [5.4]$$

where $\omega_m$ is a weight assigned to marker $m$. The linear predictor can then be written as:

$$\eta_i = \beta_0 + \sum_{j=1}^{p} \alpha_j z_{ij} + \beta_1 C_i$$

To calculate a $p$-value for a set of $M$ markers, a specific $Q_{burden}$ test statistic is computed and compared to a 1 d.o.f. $\chi^2$ distribution:

$$Q_{burden} = \left[ \sum_{m=1}^{M} \omega_m S_m \right]^2$$

This model is flexible in that it is possible to assign different weights to the markers or define the genetic score $C_i$ to account for different assumptions about the disease mechanism. For example, the Cohort Allelic Sum Test (CAST (Morgenthaler and Thilly 2007)) assumes that the presence of any rare variant increases the risk of disease and defines the genetic score $C_i = 0$ if there are no minor alleles in a region and $C_i = 1$. In addition, in order to focus on rarer variants, $\omega_m = 1$ can be defined when the MAF is below a predefined threshold and $\omega_m = 0$. It is also possible to use a continuous weighting function to weight the rare variants, with a weight of $\omega_m = 1/\sqrt{\text{MAF}_m(1 - \text{MAF}_m)}$, as proposed by Madsen and Browning (2009).

However, *burden tests* rely on the strong assumption that all rare variants in a set are causal and associated with a phenotype in the same way (same direction of association and same magnitude), which can lead to substantial loss of power if these assumptions turn out to be false (Lee et al. 2014).

### 5.1.6.3. Sequence kernel association test

The SKAT (*Sequence Kernel Association Test*) model utilizes the same statistical framework as the logistic regression and burden tests, but instead of testing the null hypothesis $H_0 : \beta_1, \ldots, \beta_M = 0$, it assumes that each coefficient $\beta_m$ follows an arbitrary distribution with zero mean and a variance equal to $\omega_m \tau$, where $\tau$ is a variance component and $\omega_m$ is a weight assigned to the marker $m$. Under this assumption, we can see that testing $H_0 : \beta_1, \ldots, \beta_M = 0$ is tantamount to testing $H_0 : \tau = 0$, which can be carried out with a *variance-component* type of test generally used in the linear mixed model (GLMM). One of the advantages of this test is that it is only necessary to model the null model and then compute the following score statistic:

$$Q_{SKAT} = \sum_{i=1}^{n} \sum_{i'=1}^{n} (y_i - \eta_i) K(\mathbf{x}_i, \mathbf{x}_{i'}) (y_{i'} - \eta_{i'})$$

where $\eta_i = \beta_0 + \sum_{j=1}^{p} \alpha_j z_{ij}$ is the linear predictor of the null model including only the covariates of the $i$th individual, and where $K(\mathbf{x}_i, \mathbf{x}_{i'}) = \sum_{m=1}^{M} \omega_m x_{im} x_{i'm}$. $K(\cdot, \cdot)$ is a kernel function and measures the genetic similarity between individuals $i$ and $i'$, weighted by a factor $\omega_m$, by way of the $M$ genetic markers of the region of interest.

This particular form of $K(\cdot, \cdot)$ is called **weighted linear kernel function**. The kernel function can take different forms to allow, for example, epistatic effects to be taken into account. In fact, any positive semidefinite function can be used as a kernel. In their article, Wu et al. (2011) propose different types of kernels that can be used in the SKAT model:

– the weighted linear kernel:

$$K(\mathbf{x}_i, \mathbf{x}_{i'}) = \sum_{m=1}^{M} \omega_m x_{im} x_{i'm}$$

implies a linear relationship between the trait of interest and the genetic variants and is equivalent to the classical logistic linear model;

– the weighted quadratic kernel:

$$K(\mathbf{x}_i, \mathbf{x}_{i'}) = (1 + \sum_{m=1}^{M} \omega_m x_{im} x_{i'm})^2$$

assumes that the model depends on the main effects and quadratic terms, as well as first-order marker-tag interactions;

– Identical By State (IBS[6]):

$$K(\mathbf{x}_i, \mathbf{x}_{i'}) = \sum_{m=1}^{M} \omega_m IBS(x_{im} x_{i'm})$$

defines the similarity between individuals as the number of identical alleles per state.

#### 5.1.6.4. *Gene-based association analysis*

In a gene-based association analysis, all variants are analyzed together within a single gene to obtain a single $p$-value representing the level of significance of the association of the entire gene with a phenotype. Unlike SNPs that have different allelic frequencies, high collinearity due to LD and different heterogeneity across human populations, a gene unit is much more stable (Neale and Sham 2004).

Furthermore, with the gene as the unit of analysis, the extension of the results to other functional analyses, such as protein–protein interactions and biological pathways, is simpler. The integration of the results of association studies with functional information can thus facilitate the understanding of biological mechanisms involved in complex diseases (Li et al. 2011).

A number of gene-based association tests have been proposed. Linear regression (for quantitative traits) and logistic regression (for binary traits), presented in sections 5.1.6.1 and 5.1.6.2, are simple methods for evaluating the overall association between a gene and a trait. On the other hand, methods using SNPs from the same gene as a analysis unit and combining test or $p$-value statistics have also been proposed.

A first method proposes to consider the largest test statistic among all the tests performed on SNPs of a single gene as the test statistic associated with that gene (Wang et al. 2007).

Nonetheless, the value of this statistic is positively correlated with the number of SNPs in the gene; although adjusting the statistic by gene size via a permutation procedure is possible, it is time consuming for large datasets (Wang et al. 2007).

Another method consists of combining the $p$-values of SNPs in a gene by a Fisher combination test (Curtis et al. 2008). However, this method assumes that the calculation of the $p$-values must be based on independent tests, which is unlikely for SNPs belonging to the same gene. Although failure to comply with this assumption is likely to increase the type-I error rate, it is nevertheless possible to use a permutation procedure to determine an empirical significance level.

---

6. A DNA segment is identical by state in two or more individuals if they have identical nucleotide sequences in that segment.

On the other hand, Pan et al. (2014) proposed an approach based on the estimation and selection of the most powerful test among a class of tests called the **optimized score sum**, covering several classes of tests such as the *burden test*, the *variance-component score test* or even the classical univariate test. The main idea is to use different values of a parameter to build weights on the SNPs, thus adapting the test to the signal sparsity and to the different directions of association between the SNPs under test.

## 5.2. Structured compression and association study

### 5.2.1. *Context*

Conventional GWAS analyses may lack the power to explain the mechanisms that take place in hereditary or multifactorial diseases. Integrating the data structure in the hypothesis testing procedure allows their power to be improved.

Meinshausen (2008) has, for example, proposed a hierarchical approach which, rather than considering the variables individually, takes into account the influence of clusters of highly correlated variables. The statistical power of this method for detecting relevant variables at a single SNP level is comparable to Bonferroni–Holm-based procedures (Holm 1979), yet with a higher level of detection as the cluster size increases, potentially inducing a higher number of false positives.

In the family of linear models, Listgarten et al. (2013) constructed a likelihood ratio test integrating confounding factors related to kinship and to the structure of the population. Based on the linear mixed model, it uses two random effects: one for capturing associations and the other for capturing confounding factors. The results emphasize better control of the type-I error, as well as increased statistical power compared to more traditional testing procedures. Other multiple linear regression methods lead to improving the power of statistical tests by taking into account linkage imbalance within the genes (Yoo et al. 2016), or to increasing the percentage of variance explained in complex traits by grouping weakly but often associated variants around large known chromosomal regions (Pare et al. 2015).

Finally, other approaches have focused on aggregating statistical summaries of SNPs belonging to the same gene as in Kwak and Pan (2016), or still on procedures for combining $p$-values as in Petersen et al. (2013). The methods presented in these studies are used on SNPs located in coding regions (or extended intronic regions such as in Petersen et al. (2013)), but can be adapted to a set of SNPs when the latter is pre-specified within a region. The statistical power of each test nonetheless remains dependent on the underlying disease model.

However, it should be noted that when too few variants are associated with a genetic trait, many variants do not cause any effect, or still that when the frequency of causal variants is too low, the statistical power of this type of approach may also be reduced, compared to simple hypotheses testing applied to individual SNPs (Lee et al. 2014). The relevance and quality of the clustering therefore play a crucial role.

### 5.2.2. *New structured compression approach*

We describe here an approach for detecting regions involved in genetic diseases as part of studies on genome-wide association. In order to reduce the size of the problem in GWAS, this new multi-marker based approach aims to construct – by compressing the additively coded SNP data into the matrix $\mathbf{X}$ – relevant genomic regions, which will then be tested in association with a phenotype.

We can summarize this method combining hierarchical classification and supervised learning in four stages. The notations presented here are partly represented in Figure 5.4:

– Step 1: determine a cluster structure ($\mathcal{G} = \{\mathcal{G}^h\}_{h=1}^H$). First of all, we determine a hierarchical structure of height $H$, where the structure $\mathcal{G}^h$ at the intermediate height $h$ is composed of $G_h$ clusters: $\{\mathcal{G}_g^h\}_{g=1}^{G_h}$. This structure is defined from an additive coding of the SNP matrix, $\mathbf{X} \in \mathbb{R}^{n \times D}$, to which a hierarchical ascending classification algorithm is applied (HAC), which is spatially constrained and takes into account linkage imbalance (Dehman et al. 2015).

– Step 2: build a matrix of aggregated SNPs ($\forall h, \mathbf{X} \overset{\mathcal{G}^h}{\leadsto} \tilde{\mathbf{X}}^h$). In a second step, we reduce the dimension of the matrix $\mathbf{X}$, by using the group structure $\mathcal{G}$ defined in the previous step in order to obtain for each height $h$, an aggregated matrix $\tilde{\mathbf{X}}^h \in \mathbb{R}^{n \times G_h}$, where by construction, $G_h < D, \forall h$.

– Step 3: estimate the optimal cut level ($h^\star : \mathbf{X} \overset{\mathcal{G}^{h^\star}}{\leadsto} \tilde{\mathbf{X}}^{h^\star}$). In a third step, we estimate the optimal cut level $h^\star$ of the hierarchy built in step 1 by means of a supervised learning algorithm combining steps 1 and 2, and define the optimal SNP matrix $\tilde{\mathbf{X}}^{h^\star} \in \mathbb{R}^{n \times G_{h^\star}}$.

– Step 4: test the relevance of the aggregated groups. Finally, in a fourth step, a multiple testing procedure makes it possible to identify significant associations between the aggregated variables of the $\tilde{\mathbf{X}}^{h^\star}$ matrix and the phenotype $\mathbf{y}$.

We now describe every step of our method. For each step, we propose two levels of description: an overview that we would like to be self-sufficient in order to master the mechanisms at play, then a focus allowing for further and deeper understanding of these mechanisms.

### 5.2.2.1. *Step 1: determine a group structure*
#### 5.2.2.1.1. Overview

In order to define a structure $\mathcal{G}$ that is able to take into account LD in the $D$ SNP matrix, we utilize the hierarchical classification algorithm *adjclust*[7], developed by Ambroise et al. (2019), which spatially constrains the definition of clusters by allowing only adjacent areas to be merged.

More precisely, it is the similarity measure, used in the *adjclust* algorithm, that is constrained on a specific band, which induces non-zero similarities only on this band.

More formally, we denote by $\mathbf{S}$ the matrix of the $D \times D$ similarities on a band of width $\ell + 1$, with $\ell \in [1, \ldots, D]$:

$$\mathbf{S} = \begin{cases} dist(i,j) & \text{if } 1 \leq i,j \leq D \\ 0 & \text{if } |i-j| \geq \ell \end{cases}$$

Here, $D$ represents the number of naturally ordered objects that should be classified, namely SNPs. This definition is not restrictive insofar as $\ell = D$ always works. Nevertheless, considering the large size of the genomic data, we have mainly focused on the case in which $\ell \ll D$.

These different elements result in improving time and spatial complexity in hierarchical classification applied to a whole genome. We should note that a similar hierarchical classification algorithm using Ward's criterion, CONISS (*CONstrained Incremental Sums of Squares*), has also been proposed in (Grimm 1987). However, the quadratic complexity of its implementation makes its utilization impracticable with large sets of genomic data.

#### 5.2.2.1.2. Focus on *adjclust*

The main characteristics of *adjclust* are as follows: (1) it is capable of computing each Ward's linkage[8] involved in the constrained hierarchical classification in constant time; (2) it stores candidate classes for clustering in a binary heap structure (*min-heap* in our case).

**Ward's linking criterion as a function of pre-computed sums**. To reduce the computational complexity of each Ward link, it can be observed that the sum of all similarities of any class $K = \{u, \ldots, v-1\}$ of size $k = v - u$ can be expressed as a sum of elements in the $\ell_k = \min(\ell, k)$ first subdiagonal of $\mathbf{S}$.

---

7. An R package is available at: https://cran.r-project.org/web/packages/adjclust.
8. More precisely, the way two classes will merge.

To see this more clearly, we define, for $1 \leq r, l \leq D$, the sum of all the elements of **S** in the first $l$ subdiagonals of the $r \times r$ upper-right block of **S** as:

$$P(r,l) = \sum_{\substack{1 \leq i,j \leq r \\ |i-j| < l}} dist(i,j)$$

and symmetrically, $\bar{P}(r,l) = P(D+1-r, l)$ $P$ and $\bar{P}$ being sums of elements in pencil-shaped areas that we call **forward pencils** and **backward pencils**, as shown in Figure 5.3.

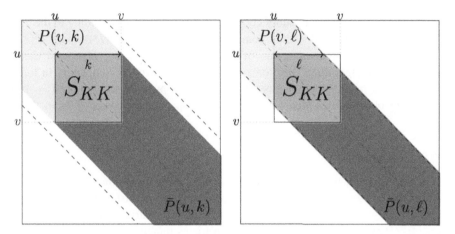

**Figure 5.3.** Example of forward pencil (in yellow and green) and backward pencil (in green and blue), with an illustration of equation [5.5] for the class K = {u, ..., v − 1}. a) The class is smaller than the band ($k \leq \ell$); b) the class $K$ is larger than the band ($k \geq \ell$)

The advantage of calculating the sums $P$ and $\bar{P}$ is to allow the calculation of any sum $S_{KK}$ of all similarities within a cluster $K$ of contiguous individuals:

$$P(v, \ell_k) + \bar{P}(u, \ell_k) = S_{KK} + P(p, \ell_k) \qquad [5.5]$$

where $\ell_k = \min(\ell, k)$ and the $P(p, \ell_k)$ are the solid pencils of width $\ell_k$.

The recursive calculation of these pencils using cumulative sums for the pre-computation step has a time complexity equal to $p\ell$ (see the proof in Dehman (2015)).

**Storage of candidate classes for clustering in a *min-heap*.** Each iteration $t$ of a hierarchical classification consists of finding the minimum number of elements $D - t$

corresponding to the candidate classes to be merged into $D - t + 1$ classes. These elements are usually stored in a sorted list, along with the algorithmic cost of deleting and inserting a linear element in $D$. In order to reduce this algorithmic complexity, *adjclust* uses a partially ordered data structure called *min-heap* (Williams 1964) to store candidate elements for clustering.

A *min-heap* is a binary tree structure constructed in such a way that the value of each node is less than the value carried by each child's node. The advantage of such a structure lies in realizing a trade-off between storing all the elements in memory and the time to find the smallest element at each iteration.

In the end this yields a memory complexity of $\mathcal{O}(D\ell)$, corresponding to the $2D\ell$ precomputed pencils, and a time complexity of $\mathcal{O}(D(\ell + \log(D))$, with $\mathcal{O}(D\ell)$ resulting from the pre-computation of the pencils and $\mathcal{O}(D\log(D))$ from the $D$ iterations of the algorithm. Therefore, *adjclust* benefits from a complexity in linear time and linear memory complexity when $\ell \ll D$. In practice, we will use $\ell = 100$, after having observed that higher values have no impact on the power of the method.

#### 5.2.2.2. *Step 2: build an aggregated SNP matrix*

#### 5.2.2.2.1. Overview

A way to work around the issues associated with high-dimensional statistics, particularly those associated with multiple tests, consists of reducing the dimensionality of the matrix $\mathbf{X} \in \mathbb{R}^{n \times D}$ by defining a matrix of reduced dimension $\tilde{\mathbf{X}} \in \mathbb{R}^{n \times G}$, where $G \ll D$, with new covariates remaining representative of the initial variables. We will in particular study an approach to constructing these new covariates from the number of clusters $G$ identified via the constrained hierarchical classification described in step 1.

While classical methods make use of the initial set of the $D$ covariates to predict a phenotype $y$, we propose to integrate a dimension reduction approach into the prediction process. We define by $\tilde{x}_i^h$, the $G_h$-dimensional vector of the new covariates for observation $i$, given a fixed hierarchy level $h : G_h \ll D$. For each cluster identified by the *adjclust* algorithm (step 1), we apply a function that leads to obtaining a single variable defined as the number of minor alleles existing in the cluster. As such, for each observation $i$ and in each class $g = \{1, \ldots, G_h\}$, the variable is defined by:

$$\tilde{x}_{ig}^h = \sum_{m \in \mathcal{G}_g^h} x_{im} \qquad [5.6]$$

It can be observed that this function [5.6] is close to function [5.4] defined for the *burden tests* (see section 5.1.6.2), where a weight $\omega_d = 1$ is assigned to each SNP, given that we are not specifically focusing on rare variants, but rather on variants presenting an MAF $\geq 5\%$. In order for the values in the different clusters to be comparable, we eliminate the impact of cluster size by centering and reducing the matrix $\tilde{\mathbf{X}}$. In the following, we will refer to this new matrix $\tilde{\mathbf{X}}$ as the matrix of **aggregated-SNP**

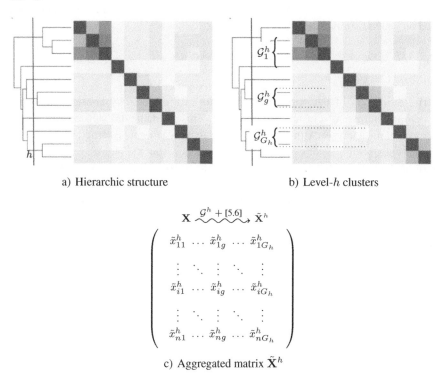

**Figure 5.4.** *Schematic view of steps 1 and 2 yielding the aggregated matrix $\tilde{\mathbf{X}}^h$ at the level $h$ of the hierarchy*

COMMENTS ON FIGURE 5.4.– *(a) Definition of the hierarchical structure $\mathcal{G}$ by the adjclust algorithm; (b) definition of the clusters at height $h$ from $\mathcal{G}^h$; (c) definition of the matrix $\tilde{\mathbf{X}}^h$ originating from the clusters $\mathcal{G}^h$. The matrices illustrating linkage disequilibria in (a) and (b) are taken from Giraud et al. (2014).*

### 5.2.2.2.2. Focus on compression types

Provided that raw data on genotype and possibly on the phenotype are available, different compressions can be taken into consideration for representing previously identified clusters of variables, including classical summaries such as the mean, median or other percentile relevant to the problem being addressed, or also the first factor(s) of a principal component analysis (Wang and Abbott 2008).

Other more sophisticated methods make use of the function coefficients of a Fourier analysis (Wang and Elston 2007) to LD weights associated with SNPs in a specific region (Li et al. 2008), or even to summaries originating from clustering based on genetic similarities (Buil et al. 2009).

### 5.2.2.3. *Step 3: estimate the optimal cut level*

#### 5.2.2.3.1. Overview

The estimation of the optimal cut level $h^\star$ in the hierarchy consists of finding the optimal number of clusters $G_h^\star$ to select. This fundamental step conditions the relevance of the analysis. Although it is known that the genome is structured by blocks of haplotypes, with little or even no recombinations within the blocks, it is not easy to determine how these blocks are distributed in the genome for a given set of SNPs.

Despite the number of studies that have been conducted to determine the optimal number of clusters in a hierarchical classification, we propose to use a supervised approach to achieve this objective. Actually, insofar as one of the purposes of GWAS is to evaluate the probability of a phenotype from genetic markers, we can utilize this information to determine the optimal number of clusters.

In order to find this *optimum*, we first propose to split our dataset into two labeled subsets, respectively, containing $n/2$ individuals: a set $\mathcal{S}_1 = (\mathbf{X}_{\mathcal{S}_1}, \mathbf{y}_{\mathcal{S}_1})$ employed to decide the cut level and another $\mathcal{S}_2 = (\mathbf{X}_{\mathcal{S}_2}, \mathbf{y}_{\mathcal{S}_2})$, and then utilized in the multiple testing procedure in step 4 to limit the inflation of type-I errors.

We can then summarize the process of choosing the cut level as follows:

1) from the set $\mathcal{S}_1$:

    a) definition of a training set: $\mathcal{T} \subset \mathcal{S}_1$;

    b) definition of a complementary validation set: $\mathcal{V} = \bar{\mathcal{T}}$;

2) from the structure $\mathcal{G}$ (step 1) and the dimension reduction function [5.6] (step 2), for each explored level $h_{min} \leq h \leq h_{max}$:

    a) definition of a training set: $\tilde{\mathcal{T}}^h = (\tilde{\mathbf{X}}_\mathcal{T}^h, \mathbf{y}_\mathcal{T})$;

    b) definition of a validation set: $\tilde{\mathcal{V}}^h = (\tilde{\mathbf{X}}_\mathcal{V}^h, \mathbf{y}_\mathcal{V})$;

    c) estimation of the coefficients of a penalized regression from $\tilde{\mathcal{T}}^h$:

$$\hat{\beta}_\mathcal{T}^h = \texttt{Penalized-Reg}(\tilde{\mathbf{X}}_\mathcal{T}^h, \mathbf{y}_\mathcal{T})$$

d) estimation of the quality of prediction $\hat{\mathbf{y}}_\mathcal{V}^h$ with respect to $\mathbf{y}_\mathcal{V}$:

$$\hat{\mathbf{y}}_\mathcal{V}^h = \tilde{\mathbf{X}}_\mathcal{V}(\hat{\boldsymbol{\beta}}_\mathcal{T}^h)^T$$
$$\hat{e}^h = \texttt{Error}(\mathbf{y}_\mathcal{V}, \hat{\mathbf{y}}_\mathcal{V}^h)$$

3) from the set of the explored levels $h_{min} \leq h \leq h_{max}$, identification of the optimal cutting level:

$$h^\star = \arg\min_h \hat{e}^h$$

Before describing the last step of our method, we can introduce several observations about the elements of this process.

REMARK. 5.1.– To speed up the searching process for the cut level, we can restrict the levels explored to an interval $[h_{min}, \ldots, h_{max}]$, which will be a priori delineated based on knowledge, for example, by assessing whether the cluster size defined at certain hierarchy levels corresponds to genomic region sizes relevant to the problem being considered.

REMARK. 5.2.–The penalized regression of the estimate described above in point (2)(c) constrains the $\ell_p \in \mathbb{N}$-norm of the coefficients, with $1 \leq p \leq 2$ in our case, to be smaller than a certain value. This choice is motivated by the fact that this type of model presents good stability properties when $p = 2$, for example. Using a more pictorial approach, small changes in the sample utilized cause little change in the estimation of the coefficients of the regression penalized by a norm $\ell_2$ (Bousquet and Elisseeff 2002). When $p = 1$, the penalty results in eliminating some variables, which may facilitate interpretation in the case of high-dimensional problems such as GWAS analyses. In practice, we will use a generalized additive model in the penalized regression described in the focus of this step, rather than the linear model.

REMARK. 5.3.–To evaluate the predictive quality of $\hat{\mathbf{y}}_\mathcal{V}^h$ in (2)(d), the mean squared error is used when $\mathbf{y}_\mathcal{V}$ is a quantitative variable, or the area under the ROC curve (AUC-ROC) when $\mathbf{y}_\mathcal{V}$ is a binary variable.

Finally, once $h^\star$ is determined, we can define from $\mathcal{G}^{h^\star}$ and the aggregation function [5.6], the optimal aggregated matrix $\tilde{\mathbf{X}}_{S_2}$ used to determine the SNP clusters in step 4 of our method.

### 5.2.2.3.2. Focus on generalized additive models

Rather than using a classical or generalized linear model[9], we prefer focusing on generalized additive models (Hastie and Tibshirani 1990) (GAM). These models – designed for modeling and describing nonlinear effects closer to real problems – present interesting properties in terms of estimation and interpretability, as we shall see in the application presented in section 5.3.

Let $(\mathbf{X}, Y)$ be a random vector. In a generalized additive model, $\mathbf{X}$ has a value in $\mathbb{R}^M$, Y belongs to a distribution of the exponential family and the linear relation is replaced by a known additive link function $\mu$:

$$\mu(\theta) = \beta_0 + \sum_{m=1}^{M} f_m(\mathbf{X}_m) \qquad [5.7]$$

where $\mu$ is monotonic and twice differentiable and $\theta = \mathbb{E}(Y = y | \mathbf{X} = \boldsymbol{x})$, with $(\boldsymbol{x}, y)$ a realization associated with the random vector.

For their part, the elements $\{f_m\}_{m=1}^{M}$ represent the variable components on a basis of (polynomial, splines, etc.) functions, and each element $f_m$ is expressed as

$$f_m(\boldsymbol{x}) = \sum_{t=1}^{T} \beta_{mt} B_{mt}(\boldsymbol{x}_m)$$

where $B_{mt}$ is the element $t$ of the basis $B$ associated with the variable $m$ and $\beta_{mt}$ is the corresponding coefficient to be estimated. In our implementation, we have used smoothing cubic splines as the basis $B$.

Finally, the regression problem penalized by a $\ell_p$ norm, for $1 \leq p \leq 2$, in its generalized additive form can be written as

$$\min_{\{f_m\}_{m=1}^{M}} \sum_{i=1}^{n} \left( y_i - \beta_0 - \sum_{m=1}^{M} f_m(x_{im}) \right)^2 + \Omega_p \left( \sum_{m=1}^{M} \lambda_m \int_{-\infty}^{+\infty} \left| f_m^{''}(x_m) \right|^2 \mathrm{d}x_m \right) \qquad [5.8]$$

---

9. This model appears in particular in step 3, making it possible to determine the optimal cut level with the penalized regression, but also in the computation of $p$-values in step 4 for identifying the relevant aggregate variables.

where $\Omega_p$ represents the penalty function applied to the elements $\{f_m\}_{m=1}^M$ of the regression and $\{\lambda_m\}_{m=1}^M \subset \mathbb{R}^+$ of the parameters regulating the compromise between the data fitting term and the penalty.

Traditionally, the estimation of parameters $\{\beta_{mt}\}_{m=1,t=1}^{M,T}$ associated with the elements $\{f_m\}_{m=1}^M$ of problem [5.8] is performed by an iterative reweighting algorithm (Penalized Iteratively Reweighted Least Squares [P-IRLS]) (see Lv and Fan (2009) Daubechies et al. (2010)) in the context of $\ell_1$-norm penalized linear models). We have personally opted for the algorithm proposed by Meier et al. (2009), adapted to GAM in high-dimensional problems.

### 5.2.2.4. *Step 4: multiple tests on aggregated SNPs*

#### 5.2.2.4.1. Overview

To test associations between aggregate variables and phenotype, we use an SMA-based analysis, as described in section 5.1.5. We nevertheless perform the computation of *p*-values on the aggregated SNP variables of the matrix $\tilde{\mathbf{X}}_{\mathcal{S}_2}^{h^*}$ determined in step 3.

For each variable $g$ aggregated of matrix $\tilde{\mathbf{X}}_{\mathcal{S}_2}^{h^*}$, a likelihood ratio test is carried out with the model:

$$\mu\left(\mathbb{E}(Y = y | \mathbf{X}_g = \boldsymbol{x}_g^{h^*})\right) = \beta_0 + f_g(\boldsymbol{x}_g^{h^*})$$
$$= \beta_0 + \sum_{t=1}^{T} \beta_{gt} B_{gt}(\boldsymbol{x}_g^{h^*})$$

by comparing the model with only the average influence $\beta_0$ with the one involving the aggregate variable $g$. We thus obtain a *p*-value according to a chi-square distribution, whose computational elements are discussed in this section.

Since we are addressing multiple hypothesis testing, an appropriate threshold should be calculated for controlling the rate of false discoveries or the global rate of false discoveries (family wise error rate (FWER)) (see Roquain (2010), and the references therein). In our high-dimensional context, it is preferable to resort to an FDR control insofar as the FWER threshold is strongly constrained by the number of tests to be performed (and therefore the problem dimension), which may result in missing relevant variables.

#### 5.2.2.4.2. Focus on the computation of *p*-values for splines

We explain here how the *p*-values associated with the aggregate variables are obtained in the case of a basis of splines. We denote by $\boldsymbol{\beta}_g \in \mathbb{R}^T$ the vector of coefficients associated with the $T$ elements of the basis for the variable $g$ and the corresponding covariance matrix $\mathbf{V}_{\beta_g}$.

For GAM, if the estimator associated with the elements of the basis is not correlated to the other terms of the model, then $\mathbb{E}(\hat{\beta}_g) = 0$, otherwise there remains a slight bias and $\mathbb{E}(\hat{\beta}_j) \simeq 0$. Under the null hypothesis $H_0 : \beta_g = 0$, we have $\hat{\beta}_g \sim \mathcal{N}(0, \mathbf{V}_{\beta_g})$, and if $\mathbf{V}_{\beta_g}$ is of full rank, then $\hat{\beta}_j^T \mathbf{V}_{\beta_g}^{-1} \hat{\beta}_g \sim \chi_T^2$.

However, the penalty involved in estimating the coefficients $\beta_g$ of problem [5.8] can lead to canceling certain elements, thus reducing the dimension of the problem and the rank of the covariance matrix. This can thus lead to working with the pseudo-inverse of the covariance matrix $\mathbf{V}_{\beta_g}^{\dagger}$ of rank $r < T$, with under the null hypothesis $\hat{\beta}_j^T \mathbf{V}_{\beta_g}^{\dagger} \hat{\beta}_g \sim \chi_r^2$.

Wood (2006) specifies that as long as the $p$-values provide a clear result, it is legitimate to make a decision therefrom. On the other hand, more caution should be exercised when approaching the rejecting threshold limit of the null hypothesis. Actually, the uncertainty in the estimation of the spline smoothing parameters was neglected in the reference distribution used for testing: these distributions – potentially with a very sharp shape depending on the degree of freedom – may assign a probability that is too low for testing statistics whose values are moderately high. In this case, to obtain more accurate $p$-values, it may be preferable to perform testing on a non-penalized model, even if it comes at a cost in terms of statistical power.

## 5.3. Application to ankylosing spondylitis (AS)

### 5.3.1. *Data*

We have evaluated the relevance of our method on data related to AS disease. The cohort used includes the French subset of the large international genetic study on AS (Groupe Française d'Etude Génétique des Spondylarthrites (GFEGS) et al. 2013). For this subset, independent cases were recruited from the rheumatology department of the Ambroise Paré Hospital (Boulogne-Billancourt, France), or by means of the patient support association (Association française des spondylarthritiques). Unrelated and matched controls in population were obtained from the center for the study of human polymorphism, or have been recruited as spouses of healthy cases. The dataset contains 408 cases and 358 controls, and each individual was genotyped for 116,513 SNPs with immunochip technology (Illumina).

To eliminate the population stratification bias in the GWAS analysis, we added the first five main genomic components – obtained by spectral decomposition of the

SNP matrix (following the EIGENSTRAT method described in Price et al. (2006)) as explanatory variables in the regression model. Since the methods evaluated here do not address missing values, we chose to impute the missing genotypes with the most frequently observed genotypic value for each SNP.

For each dataset, we filtered the values to ensure that only SNPs with a minor allelic frequency (MAF) greater than 5% were retained.

### 5.3.2. *Predictive power evaluation*

To evaluate the contribution of GAM in a GWAS study, we compare the results, in terms of predictive power, of four regression models used to estimate an optimal number of groups, namely, the cut level of the hierarchical classification. The comparison is performed based on the AUC-ROC curves obtained from the algorithm described in section 5.2.2. We respectively use *Lasso*, *group-Lasso*, *ridge* and HGAM regressions as learning methods. The results are presented in Figure 5.5. It should be noted that for *group-Lasso*, the algorithm was applied at the SNP level rather than at the aggregate SNP level.

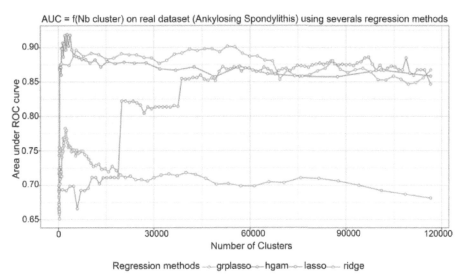

**Figure 5.5.** *AUC-ROC curves illustrating the predictive power of four statistical learning approaches for multiple group structures $\mathcal{G}^h$ – defined at different hierarchical classification cut levels – with case–control SNP data of ankylosing spondylitis*

From these curves, we observe a close behavior between the *ridge* and HGAM regressions, with a similar optimal number of groups identified. The use of cubic smoothing splines in the HGAM model significantly increases the predictive power compared to other models. The *group-Lasso* regression exhibits a globally correct predictive power, but still significantly lower than that of the HGAM regression when we fit the model to the SNP aggregates obtained at the best cutting level.

### 5.3.3. Manhattan diagram

The best cut level identified using a high-dimensional additive model is 2,750 aggregated SNPs. In the first instance, for each of these aggregated SNPs, an additive univariate model using smoothing cubic splines is estimated. In a second step, we compute the $p$-values as described in section 5.2.2.4. Figure 5.6 shows a result where 23 aggregate SNPs have been identified as significant.

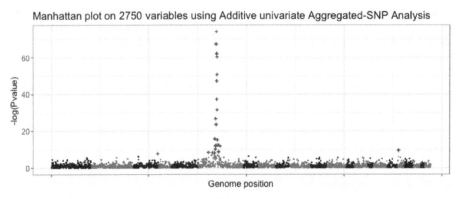

**Figure 5.6.** *Manhattan diagram of $p$-values computed for 2,750 SNPs aggregated by smoothing cubic splines*

### 5.3.4. Estimation for the most significant SNP aggregates

The curves in Figure 5.7 represent estimates of a HGAM regression obtained from the 23 most significant aggregate SNPs. These aggregate SNPs are almost all located on the same region of chromosome 6, previously identified as the location of the genetic risk factor for the disease in previous studies (Groupe Française d'Etude Génétique des Spondylarthrites (GFEGS) et al. 2013).

It is interesting to note that the significant regions identified on this chromosome explain the phenotype in a nonlinear manner. Although this complex form of phenotype–genotype linkage does not allow the detection of new associations, a significant improvement in predictive power can however be observed.

On the other hand, on chromosome 18, we can observe a new signal, of linear form, which deserves a more detailed examination based on current knowledge of the pathology.

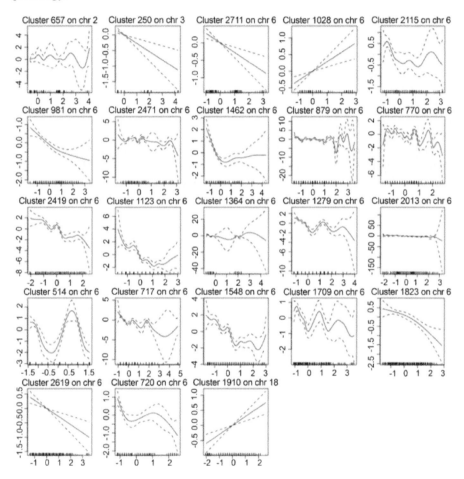

**Figure 5.7.** *Prediction representation by smoothing spline according to the values of the 23 aggregate SNPs identified as the most significantly associated with the phenotype*

## 5.4. Conclusion

We have proposed a method explicitly designed to take into account LD in GWAS. On the one hand, it combines an unsupervised approach for defining clusters of correlated SNPs by means of a dendogram, and, on the other hand, an approach for finding the optimal number of clusters, that is, the optimal cut level in the dendogram.

The fact that the LD structure has been taken into account for constructing aggregated variables representative of strongly correlated SNPs is an interesting alternative to standard marker analysis in the context of GWAS, in which it results in reducing the dimension of the initial problem. Moreover, this aggregation, coupled with a generalized additive regression model, yields significant improvement in terms of predictive power compared to other regression models, especially when the LD structure is sufficiently pronounced.

Furthermore, the application of this method to real data obtained from a study on AS has helped to reinforce these results. In addition, the use of GAM has made it possible to detect, in the HLA-specific region of chromosome 6, nonlinear behaviors between some aggregated SNPs and phenotypes that deserve further examination in light of the current knowledge on the disease.

## 5.5. References

Ambroise, C., Dehman, A., Neuvial, P., Rigaill, G., Vialaneix, N. (2019). Adjacency-constrained hierarchical clustering of a band similarity matrix with application to genomics. *Algorithms for Molecular Biology*, 14(1), 22.

Armitage, P. (1955). Tests for linear trends in proportions and frequencies. *Biometrics*, 11(3), 375–386.

Asimit, J.L., Day-Williams, A.G., Morris, A.P., Zeggini, E. (2012). ARIEL and AMELIA: Testing for an accumulation of rare variants using next-generation sequencing data. *Human Heredity*, 73(2), 84–94.

Balding, D.J., Bishop, M., Cannings, C. (2008). *Handbook of Statistical Genetics*. John Wiley & Sons, Chichester.

Bousquet, O. and Elisseeff, A. (2002). Stability and generalization. *Journal of Machine Learning Research*, 2, 499–526.

Brown, T.A. (2007). *Genomes 3*. Garland Science Publishing, New York and London.

Buil, A., Martinez-Perez, A., Perera-Lluna, A., Rib, L., Caminal, P., Soria, J.M. (2009). A new gene-based association test for genome-wide association studies. *BMC Proceedings*, BioMed Central, 3, 1–5.

Bush, W.S. and Moore, J.H. (2012). Genome-wide association studies. *PLoS Computational Biology*, 8(12), e1002822.

Chneiweis, H. (2007). Chroniques génomiques : des puces ADN en clinique ? *Médecine/Sciences*, 23(2), 210–214.

Clarke, G.M., Anderson, C.A., Pettersson, F.H., Cardon, L.R., Morris, A.P., Zondervan, K.T. (2011). Basic statistical analysis in genetic case-control studies. *Nature Protocols*, 6(2), 121.

Cochran, W.G. (1954). Some methods for strengthening the common $\chi_2$ tests. *Biometrics*, 10(4), 417–451.

Curtis, D., Vine, A.E., Knight, J. (2008). A simple method for assessing the strength of evidence for association at the level of the whole gene. *Advances and Applications in Bioinformatics and Chemistry: AABC*, 1, 115.

Daubechies, I., DeVore, R., Fornasier, M., Güntük, C.S. (2010). Iteratively reweighted least squares minimization for sparse recovery. *Communications on Pure and Applied Mathematics*, 63(1), 1–38.

Dehman, A. (2015). Spatial clustering of linkage disequilibrium blocks for genome-wide association studies. PhD Thesis, Université d'Évry Val d'Essonne; Université Paris-Saclay; Laboratoire de Mathématiques et Modélisation d'Evry.

Dehman, A., Ambroise, C., Neuvial, P. (2015). Performance of a blockwise approach in variable selection using linkage disequilibrium information. *BMC Bioinformatics*, 16, 148.

Dudbridge, F. and Gusnanto, A. (2008). Estimation of significance thresholds for genomewide association scans. *Genetic Epidemiology*, 32(3), 227–234.

Excoffier, L. and Slatkin, M. (1995). Maximum-likelihood estimation of molecular haplotype frequencies in a diploid population. *Molecular Biology and Evolution*, 12(5), 921–927.

Grimm, E.C. (1987). CONISS: A FORTRAN 77 program for stratigraphically constrained cluster analysis by the method of incremental sum of squares. *Computers & Geosciences*, 13(1), 13–35.

Groupe Française d'Etude Génétique des Spondylarthrites (GFEGS), Nord-Trøndelag Health Study (HUNT), Spondyloarthritis Research Consortium of Canada (SPARCC), Wellcome Trust Case Control Consortium 2 (WTCCC2), Bowness, P., Gafney, K., Gaston, H., Gladman, D.D., Rahman, P., Maksymowych, W.P. (2013). Identification of multiple risk variants for ankylosing spondylitis through high-density genotyping of immune-related loci. *Nature Genetics*, 45(7), 730–738.

Guinot, F., Szafranski, M., Ambroise, C., Samson, F. (2018). Learning the optimal scale for GWAS through hierarchical SNP aggregation. *BMC Bioinformatics*, 19(1), 459.

Hastie, T.J. and Tibshirani, R.J. (1990). *Generalized Additive Models*. CRC Press, Boca Raton, FL.

Holm, S. (1979). A simple sequentially rejective multiple test procedure. *Scandinavian Journal of Statistics*, 6(2), 65–70.

Kwak, I.-Y. and Pan, W. (2016). Adaptive gene- and pathway-trait association testing with GWAS summary statistics. *Bioinformatics*, 32, 1178–1184.

Lee, S., Wu, M.C., Lin, X. (2012). Optimal tests for rare variant effects in sequencing association studies. *Biostatistics*, 13(4), 762–775.

Lee, S., Abecasis, G.R., Boehnke, M., Lin, X. (2014). Rare-variant association analysis: Study designs and statistical tests. *American Journal of Human Genetics*, 95(1), 5–23.

Li, B. and Leal, S.M. (2008). Methods for detecting associations with rare variants for common diseases: Application to analysis of sequence data. *The American Journal of Human Genetics*, 83(3), 311–321.

Li, M., Wang, K., Grant, S.F.A., Hakonarson, H., Li, C. (2008). ATOM: A powerful gene-based association test by combining optimally weighted markers. *Bioinformatics*, 25(4), 497–503.

Li, M.-X., Gui, H.-S., Kwan, J.S.H., Sham, P.C. (2011). Gates: A rapid and powerful gene-based association test using extended simes procedure. *The American Journal of Human Genetics*, 88(3), 283–293.

Listgarten, J., Lippert, C., Kang, E.Y., Xiang, J., Kadie, C.M., Heckerman, D. (2013). A powerful and efficient set test for genetic markers that handles confounders. *Bioinformatics*, 29(12), 1526–1533.

Lv, J. and Fan, Y. (2009). A unified approach to model selection and sparse recovery using regularized least squares. *The Annals of Statistics*, 37(6A), 3498–3528.

Madsen, B.E. and Browning, S.R. (2009). A groupwise association test for rare mutations using a weighted sum statistic. *PLoS Genetics*, 5(2), e1000384.

Maher, B. (2008). Personal genomes: The case of the missing heritability. *Nature News*, 456(7218), 18–21.

Manolio, T.A., Collins, F.S., Cox, N.J., Goldstein, D.B., Hindorff, L.A., Hunter, D.J., McCarthy, M.I., Ramos, E.M., Cardon, L.R., Chakravarti, A. et al. (2009), Finding the missing heritability of complex diseases. *Nature*, 461(7265), 747–753.

Meier, L., Geer, S.V.D., Buhlmann, P. (2009). High-dimensional additive modeling. *The Annals of Statistics*, 37, 3779–3821.

Meinshausen, N. (2008). Hierarchical testing of variable importance. *Biometrika*, 95(2), 265–278.

Mendel, G.J. (1865). Experiments concerning plant hybrids. *Proceedings of the Natural History Society of Brünn*, IV, 3–47.

Morgenthaler, S. and Thilly, W.G. (2007). A strategy to discover genes that carry multi-allelic or mono-allelic risk for common diseases: A cohort allelic sums test (CAST). *Mutation Research/Fundamental and Molecular Mechanisms of Mutagenesis*, 615(1), 28–56.

Neale, B.M. and Sham, P.C. (2004). The future of association studies: Gene-based analysis and replication. *The American Journal of Human Genetics*, 75(3), 353–362.

Pan, W., Kim, J., Zhang, Y., Shen, X., Wei, P. (2014). A powerful and adaptive association test for rare variants. *Genetics*, 197(4), 1081–1095.

Paré, G., Asma, S., Deng, W.Q. (2015). Contribution of large region joint associations to complex traits genetics. *PLoS Genetics*, 11(4), e1005103.

Petersen, A., Alvarez, C., DeClaire, S., Tintle, N.L. (2013). Assessing methods for assigning SNPs to genes in gene-based tests of association using common variants. *PLoS One*, 8(5), e62161.

Price, A.L., Patterson, N.J., Plenge, R.M., Weinblatt, M.E., Shadick, N.A., Reich, D. (2006). Principal components analysis corrects for stratification in genome-wide association studies. *Nature Genetics*, 38, 904–909.

Roquain, E. (2010). Type I error rate control for testing many hypotheses: A survey with proofs. *Journal de la société française de statistique*, 152(2), 23–38.

Stanislas, V., Dalmasso, C., Ambroise, C. (2017). Eigen-epistasis for detecting gene-gene interactions. *BMC Bioinformatics*, 18(1), 54.

Stephens, M., Smith, N.J., Donnelly, P. (2001). A new statistical method for haplotype reconstruction from population data. *The American Journal of Human Genetics*, 68(4), 978–989.

Sturtevant, A.H. (2001). *A History of Genetics*. Cold Spring Harbor Laboratory Press, New York.

Wang, K. and Abbott, D. (2008). A principal components regression approach to multilocus genetic association studies. *Genetic Epidemiology: The Official Publication of the International Genetic Epidemiology Society*, 32(2), 108–118.

Wang, T. and Elston, R.C. (2007). Improved power by use of a weighted score test for linkage disequilibrium mapping. *The American Journal of Human Genetics*, 80(2), 353–360.

Wang, K., Li, M., Bucan, M. (2007). Pathway-based approaches for analysis of genomewide association studies. *The American Journal of Human Genetics*, 81(6), 1278–1283.

Wetterstrand, K. (2016). DNA sequencing costs: Data from the NHGRI Genome Sequencing Program [Online]. Available at: www.genome.gov/sequencingcostsdata.

Williams, J.W.J. (1964). Algorithm 232: Heapsort. *Communications of the ACM*, 7(6), 347–348.

Wood, S.N. (2006). *Generalized Additive Models: An Introduction with R*. Chapman & Hall/CRC, Boca Raton, FL.

Wu, M.C., Lee, S., Cai, T., Li, Y., Boehnke, M., Lin, X. (2011). Rare-variant association testing for sequencing data with the sequence kernel association test. *American Journal of Human Genetics*, 89(1), 82–93.

Yoo, Y.J., Sun, L., Poirier, J.G., Paterson, A.D., Bull, B.S. (2016). Multiple linear combination (MLC) regression tests for common variants adapted to linkage disequilibrium structure. *Genetic Epidemiology*, 41, 108–121.

# 6
# Kernels for Omics

### Jérôme MARIETTE and Nathalie VIALANEIX
*Université de Toulouse, INRAE, MIAT, Castanet-Tolosan, France*

The development of high-throughput sequencing techniques is generating fast-growing volumes of data at relatively low costs. These data are often highly dimensional, heterogeneous and measured in a matched manner at several layers of the living organism.

In the context of systems biology, many methods have been developed to integrate this information, namely to combine the different views obtained from the same samples with a priori information and better understand the underlying mechanisms or better predict a quantity (often a phenotype) of interest. Out of these approaches, **kernels** have many advantages that explain why they are a commonly used approach for these applications.

In this chapter, we present the general setting of kernel-based approaches and their usefulness in the analysis of various types of biological data. In particular, we will focus on (unsupervised) exploratory approaches and data integration. We will illustrate the presented approaches on part of the data from the Tara Océan project (Karsenti et al. 2011; Bork et al. 2015) using the R package mixKernel (Mariette and Villa-Vialaneix 2018).

---

For a color version of all figures in this chapter, see: http://www.iste.co.uk/froidevaux/biologicaldata.zip.

## 6.1. Introduction

Advances in new sequencing techniques and the diversification in protocols now make it possible to study an organism at different biological scales. The latter are used to study the genome – the set of all the genes of an organism – and the transcriptome – the set of the transcribed RNAs – as well as the epigenome – the set of the molecular mechanisms that modulate the expression. The data thus produced are part of what is referred to as the "omics", in which, for example, data describing the proteome – the entire set of expressed proteins – or the metabolome – the set of metabolites, can also be found. They are heterogeneous, massive, high-dimensional and often obtained from a small number of individuals or experiments compared to the number of measured variables. Often, their characteristics make them ill-suited to conventional statistical tools. Typically, they are counts for which the usual Euclidean distance is not very informative, making them unsuitable for approaches using Gaussian distribution assumptions. In addition, they are often collected through several types of experiments in a matched manner, with the aim of combining the information provided by layer. This integration frequently requires the consideration of the combination of different types of data (continuous data, count data, spectra or functions, networks, etc.), especially when it is relevant to incorporate additional biological information into the analysis provided a priori (gene function, annotation, protein spatial structure, regulatory information between genes, etc.).

In this context, many integrative methods have been developed in order to combine this information and thus consider biological systems as a whole. To address these issues, **kernel**-based approaches are commonly used (Schölkopf et al. 2004) because they offer a natural framework for the analysis of these data. By working on similarity measures between samples (whose number is generally low), these approaches have proven to be more computationally and memory efficient than approaches based on the classical representation of data in individuals × variables tables, because they compress the information contained in a number of measured variables. In addition, **kernels** provide a justified mathematical framework that makes it possible to extend many standard statistical approaches in a natural way. Finally, they are adapted to very diverse types of data and provide a common framework for combining these data, whether for exploratory (Rappoport and Shamir 2018) or predictive purposes (Lanckriet et al. 2004; Borgwardt et al. 2005).

In this chapter, we present these approaches and their usefulness to various types of biological data. More specifically, section 6.2 defines what is referred to as a **kernel** and presents the notion in the wider context of **relational data**. Section 6.3 describes extensions of classical methods to the kernel framework by focusing on exploratory, unsupervised approaches. The reader is referred to Ben-Hur et al. (2008) and Vert (2007) for a description of the supervised kernel approaches in biology.

Then, section 6.4 describes the approaches used for data integration with kernels as part of the exploratory analysis, by describing the approach proposed in Mariette and Villa-Vialaneix (2018) more specifically. Finally, section 6.5 illustrates the implementation of these methods for the analysis of a part of the Tara Ocean project data (Karsenti et al. 2011; Bork et al. 2015) using the R package mixKernel available on CRAN[1].

## 6.2. Relational data

In this chapter, we consider that the samples of interest (also called observations) are denoted by $(x_i)_{i=1,...,n}$ and that they take their values in an arbitrary space $\mathcal{X}$, which covers all the examples discussed in the introduction.

### 6.2.1. Data described by the kernel

A **kernel** is a function $K : \mathcal{X} \times \mathcal{X} \to \mathbb{R}$, which can be evaluated for any pair of observations: $k_{ii'} := K(x_i, x_{i'})$. This function measures a similarity between pairs of observations and it is symmetric ($\forall\, x, x' \in \mathcal{X}$, $K(x, x') = K(x', x)$) and positive ($\forall\, N \in \mathbb{N}$, $\forall\, \{\alpha_i\}_{i=1,...,N} \subset \mathbb{R}$, and $\forall\, \{x_i\}_{i=1,...,N} \subset \mathcal{X}$, $\sum_{i,i'=1}^{N} \alpha_i \alpha_{i'} K(x_i, x_{i'}) \geq 0$). It is equivalent to the definition of a **kernel matrix**, $\mathbf{K} := (k_{ii'})_{i,i'=1,...,n}$, of dimensions $n \times n$, which is therefore symmetric positive semidefinite.

> The purpose of this box is to illustrate the diversity of kernels, with particular focus on data originating from biology, by proposing a number of examples using data of various forms (numerical, sequences, graphs, etc.) and origins (transcriptomic data, proteins, metabolites, etc.). It does not intend to be exhaustive.
>
> EXAMPLE 6.1.– *Kernel for numerical data. One of the most frequently used kernels is the* **Gaussian kernel**: $K(x, x') = exp\left(-\gamma \|x - x'\|_{\mathbb{R}^p}^2\right)$, *for $\gamma > 0$. However, this kernel is more reliable for describing continuous data with moderate distribution skewness. In order to use it with count data from high-throughput sequencing techniques (such as RNA-seq data, for example), it is common practice to perform a preprocessing to transform the initial data into continuous data with reduced variability. The most frequent transformations are the logarithmic transformation or centering and scaling to unit variance by gene (Gönen and Margolin 2014).*
>
> Gaussian kernels (or other kernels for numerical data such as the polynomial kernel or even the linear kernel, which corresponds to the regular dot product) are frequently used to process samples described by non-numerical data. There exist numerous applications for the precomputation of numerical descriptors for the identification of splice sites from DNA

---

1. Available at: https-2pt://cran.r-project.org/package=mixKernel.

sequences as in Meher et al. (2016), or for protein classification based on various numerical descriptors (Qiu et al. 2007).

EXAMPLE 6.2.– *Kernels for Sequences. Many kernels have been proposed to compute similarities between biological sequences, namely, sequences $x_i = (x_{i1}, ..., x_{ip})$ that have their values in $\mathcal{X} = \mathcal{A}^{\otimes p}$, where $\mathcal{A}$ is a finite alphabet. Such sequences can, for example, be DNA molecules (with $\mathcal{A} = \{A, C, T, G\}$) or proteins (with $\mathcal{A}$ as the set of amino acids). For protein sequences, Jaakkola et al. (2000) define a kernel based on a hidden Markov chain model that corresponds to the family of proteins of interest and then use the Gaussian kernel of the Fisher score between two proteins based on this model. Leslie et al. (2002, 2004) proposed alternative kernels that are based on the occurrences of k-mers in the sequence: this approach is called* **spectrum kernel**. *Alternative approaches for defining kernels on sequences utilize alignment or local alignment-based methods, such as the general* **convolution kernel** *approach presented in Haussler (1999), which is extended for the definition of a kernel between proteins in Saigo et al. (2004). An alternative based on the principle of alignment is also described in Qiu et al. (2007). Here, the MAMMOTH algorithm, for structural alignment of protein sequences (Oritz et al. 2002), produces an asymmetric score between pairs of sequences. This score is then transformed into a kernel by the approach described in Tsuda (1999).*

EXAMPLE 6.3.– *Kernels for Structured Data: Trees, Graphs. Many applications in biology use data represented by structured objects such as trees or graphs: phylogenetic trees, fragmentation trees derived from spectra obtained in metabolomics by mass spectrometry, coexpression graphs between genes, protein/protein interaction graph, representation of a 3D protein structure in the form of a graph, etc. These objects are also suitable for kernel-based processing approaches.*

Tree kernels either utilize approaches relying on inter-vertex comparisons, namely, approaches that take advantage of the tree structure by comparing sub-trees or paths commonly shared by both trees (see Shen et al. (2014); Dührkop et al. (2015); Brouard et al. (2016) for several examples corresponding to trees built with mass spectral fragmentation trees in metabolomics; see also Vert (2002), which proposes a kernel for comparing phylogenetic trees based on the combination of a probabilistic evolution model and similarity between subtrees).

Concerning the graphs, we should differentiate kernels between the vertices of a given graph, often based on regularizations of the graph Laplacian (Kondor and Lafferty 2002; Smola and Kondor 2003), from kernels between graphs, based on substructure comparisons (Ramon and Gärtner 2003; Mahé and Vert 2009) or path comparisons (Gärtner et al. 2003; Vishwanathan et al. 2010). Vertex kernels are used (e.g. Vert and Kanehisa (2003); Rapaport et al. (2007)) to link gene expression and metabolic pathways. Graph kernels are used in Borgwardt et al. (2005) to compare proteins on the basis of their 3D structure and in Mahé and Vert (2009) to compare any molecule on the basis of their covalent lattice between atoms.

**Box 6.1.** *Kernel examples*

This formalism makes it possible to represent the arbitrary space $\mathcal{X}$ as a space endowed with a standard distance and dot product. In effect, Aronszajn (1950) shows that, when the definition conditions for the kernel are fulfilled, it defines a unique Hilbert space $\mathcal{H}$, called a "feature space" that is endowed with the dot product $\langle .,.\rangle_\mathcal{H}$, and a unique feature map $\phi : \mathcal{X} \to \mathcal{H}$, linked to the kernel by the relation:

$$\forall\, x, x' \in \mathcal{X}, \qquad K(x, x') = \langle \phi(x), \phi(x') \rangle_\mathcal{H}. \qquad [6.1]$$

In practice, $\mathcal{H}$ and $\phi$ are not explicit, but are used implicitly through the kernel. This is called the **kernel trick**. It consists of using the embedding of $\mathcal{X}$ in $\mathcal{H}$ by expressing the dot products and distances in $\mathcal{H}$ from the kernel values. For example, a distance $d_\mathcal{H}(x, x')$ is defined between two elements $x$ and $x'$ in $\mathcal{X}$, as the distance between their respective images in $\mathcal{H}$, $\|\phi(x) - \phi(x')\|_\mathcal{H}$, which using equation [6.1], can be written as:

$$d_\mathcal{H}(x, x') = \sqrt{K(x, x) + K(x', x') - 2K(x, x')}. \qquad [6.2]$$

The use of the kernel trick in data mining and learning is described in more detail in section 6.3.

### 6.2.2. Data described by a general (dis)similarity measure

As described in section 6.2.1, the case of data described by a kernel provides a rigorous framework for the analysis of heterogeneous data. It is also strictly equivalent to the case of Euclidean data represented by their distances (implicitly calculated in the feature space). However, it is not enough to cover all applications of relational data. Actually, as described in Schleif and Tino (2015), relational data can be described by similarity or dissimilarity measures that may be outside the scope of the Euclidean framework. We will then speak of **similarity** or **dissimilarity**, whose formal definition is not completely established in the literature, but which we will formalize in the following manner:

– A **dissimilarity** is a measure $\delta : \mathcal{X} \times \mathcal{X} \to \mathbb{R}^+$ that can be evaluated for any pair of observations: $\forall\, x_i, x_{i'} \in \mathcal{X}$, $\delta_{ii'} := \delta(x_i, x_{i'})$. It is further assumed that $\delta(x, x) = 0$ and $\delta(x, x') \geq 0$ for all $x, x' \in \mathcal{X}$ and that the function is symmetric ($\forall\, x, x' \in \mathcal{X}$, $\delta(x, x') = \delta(x', x)$). The dissimilarity matrix of the observations can also be defined as $\Delta := (\delta_{ii'})_{i,i'=1,\ldots,n}$, which is a symmetric matrix of dimension $n \times n$, with zero diagonal and non-negative elements.

– A **similarity** is a measure $S : \mathcal{X} \times \mathcal{X} \to \mathbb{R}$ that can be evaluated for any pair of observations: $\forall\, x_i, x_{i'} \in \mathcal{X}$, $s_{ii'} := S(x_i, x_{i'})$. It is further assumed that this function is symmetric ($\forall\, x, x' \in \mathcal{X}$, $S(x, x') = S(x', x)$), with non-negative diagonal

($\forall x \in \mathcal{X}$, $S(x,x) \geq 0$). We can then define the similarity matrix of observations as $\mathbf{S} := (s_{ii'})_{i,i'=1,\ldots,n}$, which is a symmetric matrix of dimensions $n \times n$, but not necessarily positive.

Both of these definitions can be included in the Euclidean setting when:

– the similarity matrix ($\mathbf{S}$) is positive definite. $\mathbf{S}$ is then simply a kernel matrix and the formalism of section 6.2.1 applies;

– the dissimilarity matrix ($\mathbf{\Delta}$) is a Euclidean distance matrix (Schoenberg 1935; Young and Householder 1938).

Schleif and Tino (2015) provide a review of the approaches for analyzing non-Euclidean similarity measures. These separate into two large families: one consists of transforming a non-Euclidean similarity into a kernel (by spectrum correction, spectral truncation or spectral inversion approaches (Chen et al. 2009), or by embedding this similarity into a Euclidean space by minimizing the distortion with the initial measurements (Kruskal 1964)). The other directly uses similarity measures and relies on a formal framework called a **pseudo-Euclidean space** (Goldfarb 1984). More specifically, if $\mathbf{S}$ is any similarity matrix, it can be shown that there exist two unique Euclidean spaces, $\mathcal{E}_+$ and $\mathcal{E}_-$, and two embeddings $\phi_+ : \mathcal{X} \to \mathcal{E}_+$ and $\phi_- : \mathcal{X} \to \mathcal{E}_-$ such that:

$$\forall\, x_i, x_{i'} \in \mathcal{X}, \qquad s_{ii'} = \langle \phi_+(x_i), \phi_+(x_{i'}) \rangle_{\mathcal{E}_+} - \langle \phi_-(x_i), \phi_-(x_{i'}) \rangle_{\mathcal{E}_-}.$$

Approaches based on this formalism employ extensions of statistical analysis methods that are similar to extensions based on the kernel trick by implicitly performing computations in the embedding spaces $\mathcal{E}_+$ and $\mathcal{E}_-$.

The dissimilarity framework may seem even more general. If we follow the dissimilarity/distance and similarity/dot product analogies, a dissimilarity matrix can be uniquely defined from a similarity matrix by reproducing the equality of equation [6.2]:

$$\delta_{ii'}^2 := s_{ii} + s_{i'i'} - 2s_{ii'}, \qquad [6.3]$$

provided, however, that the right-hand side of the equality is non-negative (it is then sometimes said that the similarity matrix is diagonally dominant).

Conversely, a dissimilarity matrix does not define a similarity in a unique way, even when relying on the previous analogy, because the values of $s_{ii}$ remain to be set in an arbitrary manner. Nonetheless, when a dissimilarity $\delta$ is given, the pseudo-Euclidean framework remains valid

$$\forall x_i, x_{i'} \in \mathcal{X}, \qquad \delta_{ij}^2 = \|\phi_+(x_i) - \phi_+(x_{i'})\|_{\mathcal{E}+}^2 - \|\phi_-(x_i) - \phi_-(x_{i'})\|_{\mathcal{E}-}^2.$$

In practice, in the following, we will use the following transformation (see Lee and Verleysen (2007)):

$$s_{ii'} := -\frac{1}{2}\left(\delta_{ii'}^2 - \frac{1}{n}\sum_{l=1}^n \delta_{il}^2 - \frac{1}{n}\sum_{l=1}^n \delta_{li'}^2 + \frac{1}{n^2}\sum_{l,l'=1}^n \delta_{ll'}^2\right),$$

which satisfies equation [6.3] to shift from a dissimilarity to a similarity.

When the similarity matrix $\mathbf{S}$ is positive definite, this similarity is a kernel that has the advantage of being centered (i.e. all observations have a zero mean in the feature space), which makes it directly usable for a kernel-based principal component analysis (PCA), for example (see section 6.3.2).

EXAMPLE 6.4.– *(Biodiversity). Biodiversity analysis leads to the study of samples that are characterized by the abundance of a set of species or taxa. To this end, various indexes have been proposed, in particular to compare samples (diversity $\beta$). The first indexes used, such as Jaccard (1912) and Sørensen (1948) indices, address both rare and abundant species in the same manner by comparing only the number of shared and unique species in the samples. If such data are given by $x_i = (x_{i1}, \ldots, x_{ip})$, with $x_{ij}$ as the number of observations of the species $j$ in the sample $i$, the Jaccard index is defined by:*

$$\delta_\mathrm{J}(x_i, x_{i'}) = \frac{\sum_{j=1}^p \left(\mathbf{1}_{\{x_{ij}>0,\, x_{i'j}=0\}} + \mathbf{1}_{\{x_{i'j}>0,\, x_{ij}=0\}}\right)}{\sum_{j=1}^p \mathbf{1}_{\{x_{ij}+x_{i'j}>0\}}}$$

*Other indexes, such as the Bray–Curtis dissimilarity, proposed by Bray and Curtis (1957), improve this first index by taking the abundance into account:*

$$\delta_\mathrm{BC}(x_i, x_{i'}) = \frac{\sum_{j=1}^p |x_{ij} - x_{i'j}|}{\sum_{j=1}^p (x_{ij} + x_{i'j})}$$

*Other approaches, such as the UniFrac distance (Lozupone and Knight 2005; Lozupone et al. 2007), the weighted UniFrac distance or the generalized UniFrac distance (Chen et al. 2012), propose computing the distance between samples by taking into account the phylogeny of the species being observed, as well as, sometimes, the rarity of certain species. The overall principle is to weight the quantities present in the calculation of the Jaccard index or the Bray–Curtis dissimilarity by the branch length of a phylogenetic tree between species.*

EXAMPLE 6.5.– *(Sequencing data (RNA-seq))*. *Witten (2011) shows that the Euclidean distance is ill-adapted to data from high-throughput sequencing, and develops a dissimilarity based on a Poisson distribution model, using the statistic of the likelihood ratio of a difference test between the sample means as a measure of the difference between the two samples.*

**Box 6.2.** *Dissimilarity examples*

## 6.3. Exploratory analysis for relational data

Most statistical analysis methods, whether exploratory or predictive models, assume that the input data are multivariate numerical data from $\mathbb{R}^p$ (for a $p \in \mathbb{N}^*$). The use of these standard statistical methods requires them to be adapted to data described by relations. When the data to be analyzed are presented in kernel form, the general principle of these adaptations relies on two main ingredients:

1) The computations of dot products and norms that are performed during the execution of the algorithm are replaced by their equivalent in the feature space $\mathcal{H}$. The advantage of this principle is that the dot product or the norm in the feature space are known only from the values of the kernel $(k_{ii'})_{i,i'=1,\ldots,n}$: they do not require the explicit definition of the feature space, which can be very complex. This ability to work within a feature space in an implicit way is the kernel trick, which was referred to in section 6.2.1.

2) When the algorithm requires the definition of new observations, which are not elements of the original set of observations (e.g. the barycenter of a class), these are expressed in the form of a linear or convex combination of the images by the feature map $\phi$ associated with the kernel of the initial data: $\sum_{i=1}^{n} \beta_i \phi(x_i)$. The distances to these new individuals are also exclusively expressed from values of the initial kernel and it is therefore not necessary to explicitly calculate them (only the coefficients $\beta_i$ are computed).

Here, we outline in detail particular extensions of this approach within the framework of exploratory analysis, particularly for unsupervised classification, PCA and self-organizing maps. The presentation considers kernels only, but it can be extended similarly to the example of data described by any dissimilarity by using an approach similar to the kernel trick, but based on the pseudo-Euclidean framework. In some cases, we will explain the differences between the two approaches. The section concludes with a discussion about the limitations of kernel-based approaches and some extensions for solving these limitations.

### 6.3.1. *Kernel clustering*

Clustering is a very common approach for data mining. Its objective is to gather similar observations into groups (or classes) with no a priori (unlike supervised classification, which seeks to learn a prediction model from classes known for some samples). The two simplest clustering methods are hierarchical (ascending) clustering (HC) and the $k$-means algorithm.

#### 6.3.1.1. *HC and kernel-based HC*

The principle of the HC algorithm is based on the iterative creation of a sequence of nested partitions. The method is initialized by the definition of a trivial partition, $\mathcal{P}_1$, where each cluster is a singleton $\{x_i\}$ with $i \in \{1, \ldots, n\}$. Then, by successively merging clusters two by two, the algorithm produces a hierarchy of partitions that ends with the trivial partition $\mathcal{P}_n$, composed of a single class in which all the individuals $\{x_i\}_{i=1,\ldots,n}$ are grouped. At every step of the algorithm, the mergers are chosen in order to minimize some quantity determined using a linkage criterion that defines a dissimilarity between distinct clusters.

This linkage criterion is generally directly set on the basis of dissimilarity measures provided between individuals and the algorithm is therefore directly adapted to these types of data. This is the case for linkage criteria known as **complete linkage** (maximum dissimilarities between the observations of the two classes), **single linkage** (minimum dissimilarities between the observations of the two classes) or **average linkage** (mean of the dissimilarities between the observations of the two classes).

However, because of its natural interpretation and its link with the $k$-means algorithm, one of the most widely used linkage criteria is initially defined for $\mathbb{R}^p$ data: this is Ward's (1963) linkage that measures the increase of intraclass inertia induced by the merging of two clusters:

$$\forall C, C' \subset \{x_i\}_{i=1,\ldots,n}, \ L(C,C') = I(C \cup C') - I(C) - I(C'),$$

with $I(C) = \frac{1}{|C|} \sum_{x_i \in C} \|x_i - \bar{x}_C\|_{\mathbb{R}^p}^2$ where $\bar{x}_C = \frac{1}{|C|} \sum_{x_i \in C} x_i$. Through the use of the general principles described above, the HC adapts itself to data described by a kernel:

– by redefining the linkage criterion through intraclass inertia in the feature space:

$$I(C) = \frac{1}{|C|} \sum_{x_i \in C} \|\phi(x_i) - \bar{x}_C\|_{\mathcal{H}}^2,$$

– and through the redefinition of the barycenter of the observations in the feature space by:

$$\bar{x}_C = \frac{1}{|C|} \sum_{x_i \in C} \phi(x_i).$$

This approach, described in Qin et al. (2003), was modified in Ambroise et al. (2018) to provide a simplified explicit formula for the definition of the linkage criterion $L(C, C')$, which makes it possible to define the kernel-based HC as in Algorithm 6.1.

ALGORITHM 6.1.– Kernel hierarchical clustering (HC) with Ward's linkage:
1) initialization: $\mathcal{P}_1 = \{\{x_1\}, \{x_2\}, \ldots, \{x_n\}\}$;
2) for $t = 1$ to $n - 1$;
3) compute the linkage criteria between all the clusters in the current partition $\mathcal{P}_t$:

$$L(C, C') = \frac{|C||C'|}{|C| + |C'|} \left( \frac{1}{|C|^2} \Sigma_{CC} + \frac{1}{|C'|^2} \Sigma_{C'C'} - \frac{2}{|C||C'|} \Sigma_{CC'} \right)$$

with $\Sigma_{CC'} = \sum_{i \in C, i' \in C'} k_{ii'}$;
4) merge the two clusters for which the linkage criterion is minimal to obtain the partition $\mathcal{P}_{t+1}$;
5) end for;
6) return $\{\mathcal{P}_1, \ldots, \mathcal{P}_n\}$.

### 6.3.1.2. *Kernel k-means*

An alternative to the HC algorithm is based on the definition of a centroid per class: this is the $k$-means approach. In its initial version, formulated for data with values in $\mathbb{R}^p$, the objective of this method is the definition of a partition that minimizes intraclass inertia for a number $P$ of classes, that has to be chosen a priori.

$$\arg\min_{C_1, \ldots, C_P} \frac{1}{n} \sum_{j=1}^{P} \sum_{x_i \in C_j} \|x_i - \bar{x}_{C_j}\|_{\mathbb{R}^p}^2.$$

The method proceeds iteratively by alternating a step of the calculation of the cluster centers ($\bar{x}_{C_j}$) with an assignment step for one observation (stochastic algorithm version) or for all the observations (*batch* version of the algorithm), which assigns an observation to the nearest cluster center. The initial configuration is usually chosen at random or derived from the results of a HC algorithm (which then enables using the **dendrogram** graphical representation as the basis for the choice of a number of classes $P$). The extension of this method to data described by a kernel is very similar to the HC extension. It is given in Algorithm 6.2.

Another extension for data described by dissimilarities was proposed in Kaufman and Rousseeuw (1987) in the form of a $k$-medoids algorithm, in which the centroids of each cluster are replaced by an observation in $\mathcal{X}$. This type of approximate solution can lead to a cumbersome implementation because it requires a discrete optimization phase in the learning set. A generalization was proposed in Rossi et al. (2007). The latter is based on the theoretic framework of the pseudo-Euclidean space presented in section 6.2.2, which is very close to the version of Algorithm 6.2. In addition, Rossi et al. (2007) proposed a parsimonious version of the representation of cluster centers and an efficient algorithm using pre-stored computations and a clever update to handle large volumes of data.

ALGORITHM 6.2.– Kernel $K$-means:
1) initialization: (randomly) set a partition of the observations data partitioning into $P$ clusters $\mathcal{P}_1 = \{C_1^1, \ldots, C_P^1\}$ ($t = 1$);
2) $f^1(x_i)$ denotes the cluster (in $\{1, \ldots, P\}$) of the observation $x_i$;
3) for $t = 1$ to $T$;
4) assign all observations to the clusters of $\mathcal{P}_t$ with the nearest center (**assignment phase**):

$$f^{t+1}(x_i) = \arg\min_{j=1,\ldots,P} \|\phi(x_i) - \bar{x}_{C_j^t}\|_\mathcal{H}^2$$

with $\bar{x}_{C_j^t} = \frac{1}{|C_j^t|} \sum_{x_l \in C_j^t} \phi(x_l)$ and the kernel trick that allows us to write:

$$\|\phi(x_i) - \bar{x}_{C_j^t}\|_\mathcal{H}^2 = k_{ii} - \frac{2}{|C_j^t|} \sum_{x_l \in C_j^t} k_{il} + \frac{1}{|C_j^t|^2} \sum_{x_l, x_{l'} \in C_j^t} k_{ll'}$$

5) redefine the clusters from the previous assignment (**representation phase**):

$$\mathcal{P}^{t+1} = \{C_1^{t+1}, \ldots, C_P^{t+1}\} \text{ with } C_j^t = \{x_i : f^{t+1}(x_i) = j\}$$

6) end for (**convergence**);
7) return $\mathcal{P}^{T+1}$.

### 6.3.2. *Kernel principal component analysis*

PCA is another very common approach in exploratory data analysis, whose extension to data described by a kernel was presented in Schölkopf et al. (1998). In its initial form, PCA is a dimension reduction method that projects data $(x_i)_{i=1,\ldots,n}$ of $\mathbb{R}^p$ into a lower dimension space $\mathbb{R}^q$, with $q < p$ or $q \ll p$. To this end, the observed variables are linearly combined to define a new set of variables linearly decorrelated from each other and called **principal components** (or **principal axes**). These components are obtained by spectral decomposition of the (empirical) variance/covariance matrix and thus correspond to the projection of the data into a subspace of dimension $q$ with the best reproduced inertia (i.e. the highest variance). The projection of the data on the principal components not only leads to a simplified graphical data representation, but also provides a simplification of the data by their representation in a lower dimensional space.

Within the framework of PCA, the data projection is linear and the minimized criterion is the least squares, but there are many existing extensions, such as nonlinear projection approaches presented in Lee and Verleysen (2007) or the use of non-quadratic losses for PCA as in Collins et al. (2001).

The extension of PCA to data described by a kernel is one of these nonlinear dimension reduction approaches. It exactly corresponds to the computation of a PCA in the feature space $\mathcal{H}$ associated with the kernel and involves the same steps:

1) The data are centered in the feature space: this amounts to modifying the feature map associated with the kernel ($\phi$) into a new feature map corresponding to the centered data and defined by:

$$\tilde{\phi}(x_i) = \phi(x_i) - \frac{1}{n} \sum_{i'=1}^{n} \phi(x_{i'})$$

We show that this feature map is associated with the centered kernel $\tilde{K}$:

$$\tilde{k}_{ii'} = \left\langle \tilde{\phi}(x_i), \tilde{\phi}(x_{i'}) \right\rangle_{\mathcal{H}} = k_{ii'} - \frac{1}{n} \sum_{l=1}^{n} (k_{il} + k_{i'l}) + \frac{1}{n^2} \sum_{l,l'=1}^{n} k_{ll'}$$

From a matrix point of view, $\tilde{K}$ is obtained by $\tilde{K} = K - \frac{1}{n}\mathbf{1}_n K - \frac{1}{n}K\mathbf{1}_n + \frac{1}{n^2}\mathbf{1}_n K \mathbf{1}_n$, in which $\mathbf{1}_n$ is a matrix of dimensions $n \times n$ whose elements are all equal to 1.

2) The spectral decomposition of the centered kernel $\tilde{K}$ is performed: the variance/covariance matrix in the feature space has entries $\langle \tilde{\phi}(x_i), \tilde{\phi}(x_{i'}) \rangle_{\mathcal{H}} = \tilde{k}_{ii'}$. We then denote by $(\beta_j)_{j=1,...,n}$ the eigenvectors of $\tilde{K}$ associated with eigenvalues $(\lambda_j)_{j=1,...,n}$, arranged in decreasing order. Without loss of generality, it can be assumed that these vectors are orthogonal with norms $1/\sqrt{\lambda_j}$. Therefore, the principal components in $\mathcal{H}$ can then be defined by:

$$a_j = \tilde{K}\beta_j = \sum_{i=1}^{n} \beta_{ji} \tilde{\phi}(x_i),$$

and it can then be shown that these principal components are orthonormal in $\mathcal{H}$.

3) The data are projected onto the first $q$ principal components: the coordinates of the projection of $\tilde{\phi}(x_i)$ onto the component $a_j$ are then $\langle \tilde{\phi}(x_i), a_j \rangle_{\tilde{\mathcal{H}}} = \sum_{i'=1}^{n} \beta_{ji'} \langle \tilde{\phi}(x_i), \tilde{\phi}(x_{i'}) \rangle_{\tilde{\mathcal{H}}} = [\tilde{\mathbf{K}} \beta_j]_i = \lambda_j \beta_{ji}$.

These coordinates are useful to obtain a representation of the samples in a low-dimensional space, highlighting their topology. However, unlike standard PCA, kernel PCA does not allow the variables to be represented, since the samples are described by their relations, through the kernel, not by standard numerical values. The principal components are then more difficult to interpret, because they are defined by their similarity to all samples and not by individual correlations with variables describing the samples. Moreover, the complexity of the singular value decomposition of the kernel is of the order of $\mathcal{O}(n^3)$, which makes kernel PCA poorly adapted to data sets containing a large number of observations ($n$) (see section 6.3.4 for a more thorough discussion on the limitations of kernel methods).

The extension of this approach to data represented by dissimilarities is called PCoA (principal correspondence analysis) or MDS (multi-dimensional scaling). It is based on the search for the principal coordinates $(a_i)_{i=1,...,n}$, which are vectors of $\mathbb{R}^d$ ($d \leq n$) minimizing a stress function, whose purpose is to preserve the distances between individuals in the projection space:

$$\text{Stress}(a_1, \ldots, a_n) = \sum_{i,i'=1}^{n} (\tilde{\delta}^2_{ii'} - a_i^\top a_{i'})^2$$

where $\tilde{\delta}^2$ corresponds to the centering of the dissimilarity $\boldsymbol{\Delta}^2$. The principal coordinates are then obtained by spectral decomposition of $\widetilde{\boldsymbol{\Delta}^2} = (\tilde{\delta}^2_{ii'})_{i,i'=1,...,n}$.

### 6.3.3. *Kernel self-organizing maps*

Self-organizing maps, also known as **Kohonen maps** (Kohonen 2001) or SOM (*self-organizing maps*), are a clustering method that makes it possible to combine data projection into a low-dimensional space with clustering. Initially inspired by biological principles, they belong to the family of artificial neural networks.

The SOM algorithm is close to the $k$-means algorithm, but unlike the latter, it assigns the observations to clusters that are organized on a grid with a distance (or a topological structure) as illustrated in Figure 6.1 (right-hand side). More precisely, a **grid** or **map** refers to a set of $U$ classes, often called **units** or **neurons**. This grid is equipped with a distance or a neighborhood relation. To simplify, the neurons can be considered as being positioned on a regular grid in $\mathbb{R}^2$ between the neurons $u$ and $u'$

is the usual distance of $\mathbb{R}^2$ between their respective positions. In its standard version, defined for data with values in $\mathbb{R}^p$, the SOM algorithm iterates two steps very similar to the steps of the $k$-means algorithm. These two steps are based on the definition of the **prototype** of unit $u$, $p_u$, with values in the data space ($\mathbb{R}^p$) that represents the unit (and thereby is similar to the observations classified in this unit). The two steps of the algorithm are then:

– the **assignment step**, in which a randomly chosen observation ($x_i$) (stochastic version) or all the observations (version *batch*), are assigned to the unit whose prototype is closest to the observation:

$$f(x_i) := \arg\min_{u=1,\ldots,U} \|x_i - p_u\|_{\mathbb{R}^p}^2$$

– the **representation step**, which recalculates the values of the prototypes to make them agree with the new clustering. In its stochastic version, this step can be seen as a pseudo-stochastic gradient descent around the observation $x_i$ processed in the assignment step, for the prototypes of the neighboring units of $f(x_i)$:

$$p_u \leftarrow p_u + \mu H(f(x_i), u)(x_i - p_u)$$

where $\mu > 0$ is generally chosen such that it decreases with the iterations ($t$) of the algorithm, for example in $1/t$, and $H$ is a decreasing function of the distance on the grid. $H$ is generally chosen to be piecewise constant, or with a Gaussian shape, and its intensity decreases during the learning process (generally to be restricted to the function which is equal to 1 if and only if $f(x_i) = u$ at the end of learning, which corresponds to final iterations similar to those of a $k$-means algorithm).

Several studies have been carried out on the theoretical properties of this algorithm and, in particular, on its convergence (see Cottrell et al. (2016) for a review). In its standard version, there is no proof of any theoretical guarantee that the SOM algorithm converges, unlike the $k$-means that converges to a local minimum of the intra-group variance criterion. However, the advantage of the SOM algorithm is its ability to produce a data typology that can be easily interpreted. The grid serves to visualize the topology of the observations in the initial space through the use of a nonlinear projection. Its adaptation to non-Euclidean data raises questions about the definition of the prototypes in the initial space (that is not necessarily vectorial), and about the calculation of the distance between an observation and a prototype (for the algorithm assignment step). For this purpose, several extensions have been proposed for the SOM: similarly to the $k$-medoids algorithm, the median SOM (Kohohen and Somervuo 1998; Kohonen and Somervuo 2002) and its variants (Ambroise and Govaert 1996; El Golli et al. 2006) choose the prototypes in the set of all observations. Nonetheless, it has been shown by Rossi (2014) that this approach raises representation and organization issues for the resulting map. Other extensions to data represented by kernels have been proposed in Graepel et al. (1998); MacDonald and Fyfe (2000); Andras (2002); Villa and Rossi (2007), and to data described by dissimilarities (relational SOM) in some previous studies (Hammer and Hasenfuss 2010; Olteanu and Villa-Vialaneix 2015) for the *batch* and stochastic versions of the algorithm. The two general principles described in the introduction of section 6.3 can also be applied here:

– redefine the distance through the distance induced by the kernel in the feature space;

– redefine the prototypes by restricting them to be linear (or convex) combinations of the images of the initial data in the feature space. These are then written as $p_u = \sum_{i=1}^{n} \beta_{ui}\phi(x_i)$ and the algorithm essentially consists of just updating the $(\beta_{ui})_{i,u}$ without explicitly computing the prototypes.

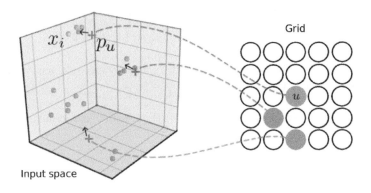

**Figure 6.1.** *Self-organizing map algorithm*

COMMENTS ON FIGURE 6.1.– *Each neuron in the grid is associated with a prototype that it represents in the origin space. An observation $x_i$ is chosen randomly. The observation $x_i$ is assigned to the winning neuron u, that is to the orange neuron, whose prototype $p_u$ is closest to $x_i$ in the initial space. The winning prototype, $p_u$, as well as the prototypes of the neighboring neurons, are updated in the direction of $x_i$ during the representation step.*

The stochastic version of the kernel SOM is given in Algorithm 6.3 and the relational version (adapted to any dissimilarities) is very close to this version, relying on equivalent computations in the pseudo-Euclidean space defined in section 6.2.

ALGORITHM 6.3.– Kernel SOM, stochastic version:
1) $\forall u = 1, \ldots, U$ and $\forall i = 1, \ldots, n$ randomly initialize $\beta_{ui}^1$ in $[0, 1]$ such that $\sum_{i=1}^n \beta_{ui}^1 = 1$;
2) and set: $p_u^1 = \sum_{i=1}^n \beta_{ui}^1 \phi(x_i)$;
3) for $t = 1$ to $T$;
4) draw a random sample: $i \in \{1, \ldots, n\}$ (**assignment step**):

$$f^{t+1}(x_i) = \underset{u=1,\ldots,U}{\arg\min} \|\phi(x_i) - p_u^t\|_{\mathcal{H}}^2$$

$$= \underset{u=1,\ldots,U}{\arg\min} \left( k_{ii} - 2\sum_{l=1}^n \beta_{ul}^t k_{il} + \sum_{l,l'=1}^n \beta_{ul}^t \beta_{ul'}^t k_{ll'} \right)$$

5) for all $u = 1, \ldots, U$ (**representation step**):

$$p_u^{t+1} = p_u^t + \mu_t H^t(f^{t+1}(x_i), u)(x_i - p_u^t)$$
$$\Leftrightarrow \beta_u^{t+1} = \beta_u^t + \mu_t H^t(f^{t+1}(x_i), u)\left(\mathbf{1}_i^n - \beta_u^t\right)$$

where $\mathbf{1}_i^n$ is a vector of dimension $n$ with a single non-zero element, the element $i$, whose value is 1;
6) end for;
7) return $(p_u^{T+1})_u$ (prototypes) and $(f^{T+1}(x_i))_i$ (the final clustering on the map).

### 6.3.4. Limitations of relational methods

Relational approaches experience two main disadvantages that may limit their utilization for statistical learning. We describe here two limitations that are frequently highlighted: their lack of interpretability and their computational complexity. Possible answers to these issues are also described.

#### 6.3.4.1. Lack of interpretability

Should the initial data be described by variables, the values of these variables would have been condensed into the kernel, and would no longer be directly usable for the interpretation of the results. In particular, for methods based on cluster centers (centers of gravity for the $k$-means algorithm or prototypes for the SOM algorithm), the representatives no longer have an explicit form that would be easily interpretable in the initial space. Indeed, they are (symbolically) represented by their proximity to the observations $\sim \sum_{i=1}^n \beta_{ui} x_i$. When $n$ is large, this type of representation makes it very difficult to interpret the meaning of clusters, contrary to the standard Euclidean setting. Similarly, for methods similar to kernel PCA, the factorial axes no longer have an explicit representation in the initial space and therefore suffer from the same lack of interpretability as the cluster representatives. To solve this difficulty, various proposals have been made:

– Description of prototypes and axes in more parsimonious formats: in this proposal, the representative of each cluster (cluster center or prototype), or the principal component in PCA, is expressed with only a small number of non-zero coefficients $\beta_{ui}$. In particular, this approach is proposed by Hofmann et al. (2015) who describe various strategies for obtaining parsimonious representations (hard thresholding, $\ell_1$ penalty, etc.) in a supervised setting. Their solutions can be easily extended to the unsupervised setting, which has been done by Rossi et al. (2007) for the $k$-means algorithm, and in Mariette et al. (2017) for the SOM algorithm.

– Permutation of the initial values: when the type of the $\{x_i\}_{i=1,\ldots,n}$ allows it, that is to say when the observations are, for example, represented by numerical variables, approaches based on permutations of the initial values can be used to assess the influence of a variable on the final result (classification, PCA projection onto the axes, etc.) and order the variables by the importance of their impact on the result of the analysis. This is the approach developed in Mariette and Villa-Vialaneix (2018) for the kernel PCA.

– Variable selection: likewise, for problems where the observations $\{x_i\}_{i=1,\ldots,n}$ are described by $p$ numerical variables, variable selection-based approaches can be incorporated into the model. They generally consist of learning a weighting $w_j \geq 0$ for each variable $j \in \{1,\ldots,p\}$ with a parsimony constraint (e.g. a $\ell_1$ penalty) on the vector $w = (w_1,\ldots,w_p)$. This type of approach has been developed for the supervised setting (see, for example, Allen (2013)) and cannot be extended to the unsupervised setting in a straightforward manner, because it is based on the optimization of the cost function associated with the regression or the classification.

### 6.3.4.2. *Large (computational) complexity as the number of samples (n) increases*

The complexity of kernel PCA is $\mathcal{O}(n^3)$ (spectral decomposition complexity) and the complexity of the kernel SOM algorithm (in its stochastic version) is $\mathcal{O}(\gamma n^3 U)$ (Rossi 2014) for a number of iterations $T \sim \gamma n$. In addition to the parsimonious description of prototypes and axes presented above, the main methods for addressing this problem are as follows:

– Nyström approximation: the Nystrom approximation (Williams and Seeger 2000) yields a low cost approximation of the spectral decomposition of a kernel $K$. More precisely, the spectral decomposition of $K$ is approximated by selecting (randomly or more efficiently, as proposed in Kumar et al. (2012)) $m$ observations among $\{x_i\}_{i=1,\ldots,n}$, $\mathcal{T}_m$, and utilizing the spectral decomposition of the reduced kernel $\mathbf{K}^{(m)} = (k_{ii'})_{i,i' \in \mathcal{T}_m}$. The total cost of the approach has a complexity dominated by $\mathcal{O}(nm^2)$ and when the kernel $\mathbf{K}$ has a rank lower than or equal to $m$, the approximation is exact.

– Exact approaches: Rossi et al. (2007) and Mariette et al. (2014) propose approaches for reducing the complexity of kernel-based approaches. These approaches rely on storing intermediate computations and updating prototypes and distance computations at a lower cost. More precisely, for a storage cost of $\mathcal{O}(nU)$, the computational cost of this approach for the SOM algorithm is $\mathcal{O}(\gamma n^2 U)$ for a number of iterations of the order of $T \sim \gamma n$.

## 6.4. Combining relational data

### 6.4.1. *Data integration in systems biology*

The development of data acquisition techniques, and in particular of high-throughput sequencing techniques, as well as the fact that omics data are becoming increasingly available, have created significant needs for the development of **data integration** methods or **multi-omics approaches**. There are numerous articles that review these approaches (Noble 2004; Kristensen et al. 2014; Franzosa et al. 2015; Ritchie et al. 2015; Bersanelli et al. 2016) and they propose typologies for these approaches according to the types of mathematical tools used or the objectives of the method.

The overall objective is to combine $M$ types of data collected from the same $n$ individuals. Each of these data types could possibly have values in a not necessarily numerical space and provide a specific view of the data. The combination can have an unsupervised, exploratory purpose (understanding the structure of the samples, achieve a data projection onto a lower dimensional space in order to visualize them, perform an unsupervised classification or a typology, as described in section 6.3) or a predictive, supervised objective (learning a prediction function using the information provided by the $M$ datasets to predict a quantity of interest from the samples). According to the nomenclature in some previous studies (Noble 2004; Ritchie et al. 2015; Rappoport and Shamir 2018), illustrated in Figure 6.2, we can then choose to:

– Concatenate the $M$ data tables into a single one on which a single (exploratory or predictive) analysis is performed. These approaches generally assume that the various data tables are numerical and use this property to obtain a relevant summary of the $M$ information before analysis, as in Singh et al. (2018) for the canonical correlation analysis.

– Perform $M$ independent analyses on each of the data tables and combine the results. These approaches are generally named ensemble methods and more specifically used for supervised analyses for which the combination of results is, for example, the average prediction. In the unsupervised setting, the combination of several results is less natural and the strategies are varied, as described by Vega-Pons and Ruiz-Schulcloper (2011) for clustering.

– Embed the $M$ data tables into a common representation space in which they are combined before a single analysis. These approaches generally employ data representations in the form of graphs or kernels and combine them into a meta-graph (Tang et al. 2009) or a meta-kernel. The advantage of these approaches, which will be the object of our focus in the rest of this section, is that they allow for the combination of heterogeneous data, namely, data that do not all have a numerical representation. In addition, they generally use a data representation in the form of relationships between samples: given that the number of samples is often small in genomics compared to the number of traits or variables describing them, the integration is faster and easier. This type of integration can be considered as being carried out at an intermediary level, between the data and the results, since the integration is generally achieved at the level of the relationships between samples according to the different views.

### 6.4.2. Kernel approaches in data integration

One of the advantages of kernel approaches is that they enable the combination of heterogeneous data sources obtained from the same individuals by proposing a uniform representation framework. This approach, known as **multiple kernel learning**, aims to combine $M$ different kernels, $(K^m)_{m=1,\ldots,M}$, obtained from the same $n$ observations, into a single kernel:

$$K^* = \sum_{m=1}^{M} \gamma_m K^m \quad \text{such that} \quad \begin{cases} \gamma_m \geq 0, \ \forall\, m = 1, \ldots, M \\ \sum_{m=1}^{M} \gamma_m = 1 \end{cases} \quad [6.4]$$

By construction, $K^*$ is also positive symmetric (and thus a valid kernel) and can therefore be used in exploratory or predictive analyses as a summary of the integrated information originating from the $M$ initial kernels.

It is natural to choose coefficients $(\gamma_m)_{m=1,\ldots,M}$ that are all equal to $1/M$. However, this choice considers all the kernels in an identical manner and does not take into account the fact that some kernels may be redundant or atypical, and as the case may be, a more relevant choice is to assign them more or less weight. This issue has been extensively addressed in supervised learning where a strategy consists of choosing the $(\gamma_m)_{m=1,\ldots,M}$ to minimize the prediction error (Gönen and Alpaydin 2011). In an unsupervised framework, such an objective does not exist. For clustering, Zhao et al. (2009) propose to choose the $(\gamma_m)_{m=1,\ldots,M}$ in order to optimize the margin between the different clusters and (Yu et al. 2012; Gönen and Margolin 2014; Huang et al. 2012) minimize the intra-class variance between the different clusters in $k$-means or *fuzzy* $c$-means approaches. However, these approaches are restricted to the unsupervised classification setting and are not valid for other types of exploratory analyses such as PCA. Within the context of PCA, Speicher and Pfeifer (2017) show that direct optimization of the values of $(\gamma_m)_{m=1,\ldots,M}$ in the PCA criterion yields a trivial solution (the selection of a single kernel, the one whose variance reproduced on the $d$ axes of interest is the greatest). They therefore propose an approach that penalizes the loss of inertia induced by the combination of kernels in a nonlinear way on each axis. This approach is also restricted to the PCA framework.

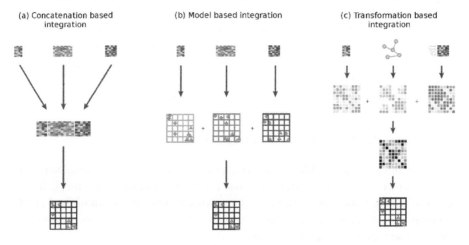

**Figure 6.2.** *Approaches to data integration*

COMMENTS ON FIGURE 6.2.– *(a) Concatenation-based integration involves the combination of datasets at the data matrix level, which can be the original data or a numerical representation; (b) model-based integration of model results requires analysis of each dataset independently before aggregating the results of the analyses; (c) transformation-based integration maps or transforms the original data in order to combine them. The combined object, which may be a graph or a kernel, can then be analyzed by any method designed to address this type of object (figure inspired from Ritchie et al. (2015)).*

Generally speaking, there are less proposals for combining kernels in an unsupervised setting and few of them address the issue of combining the kernels in a generic way, without resorting to a criteria based on a specific exploratory method (e.g. clustering or dimension reduction). Zhuang et al. (2011) identify two main criteria for a generic unsupervised combination of kernels, all computed from data $(x_i)_{i=1,\ldots,n}$ with values in $\mathbb{R}^p$:

– minimizing the reconstruction error between the meta-kernel and the initial data:

$$\|x_i - \sum_{i'=1}^{n} k_{ii'}^* x_{i'}\|_{\mathbb{R}^p}^2$$

– preserving the initial distance structure between samples in $\mathbb{R}^d$ in the meta-kernel, which translates into the minimization of the criterion $\sum_{i'=1}^{n} k_{ii'}^* \|x_i - x_{i'}\|_{\mathbb{R}^p}^2$.

Approaches using these criteria are based on the fact that the data enabling the computation of the kernels are $\mathbb{R}^p$ data and they are therefore essentially related to the representation of the samples in this space: this criterion is therefore counterproductive if we consider that the kernels obtained from the initial data are more informative than the Euclidean distance. Lin et al. (2010) and Speicher and Pfeifer (2015) use an approach that employs the preservation criterion of the local distance structure without explicitly using the values $\|x_i - x_{i'}\|_{\mathbb{R}^p}^2$. To this end, they perform a preliminary data projection (from data in $\mathbb{R}^p$) into a low-dimensional space (using, in particular, projection approaches that preserve the neighborhood such as the one described in He and Niyogi (2003)) and use the nearest neighbor graph in this projection space to replace the values $\|x_i - x_{i'}\|_{\mathbb{R}^p}^2$ by similarities based on this graph. Even if the criterion is not explicitly based on the Euclidean distance, it is however, directly derived therefrom. In addition to the fact that this method does not apply when the original data are not numerical or when the $M$ initial data do have their values in a common space, the fact that it utilizes the Euclidean distance as a measure of an underlying truth again mitigates the scope of using a kernel (different from the usual distance). Finally, Wang et al. (2017) propose a multikernel integration approach that simultaneously learns weights for the different kernels and a global similarity matrix of low rank, by successive iterations. This approach is used to combine Gaussian kernels with different parameters for the visualization of single-cell RNA-seq data and makes the underlying assumption that the data are structured into $P$ classes where $P$ is a parameter to be set.

In Mariette and Villa-Vialaneix (2018), we propose three approaches with the aim of addressing the issue of combining multiple kernels in an unsupervised setting and without a priori structure assumption. The first one leads to obtaining a consensus kernel, which is the closest kernel, on average, compared to all of the other kernels. The second and third approaches use a different point of view, similar to that of Zhuang et al. (2011), and define a kernel that minimizes the distortion with all the other kernels. The latter two approaches differ in the fact that they use a $\ell_2$ or $\ell_1$ penalty in the optimized criterion; the $\ell_1$ penalty allows a selection among the $M$

input kernels. These approaches are implemented in the R package mixKernel and illustrated in section 6.5.

### 6.4.3. *A consensual kernel*

The first proposal, STATIS-UMKL, uses a STATIS-type approach (L'Hermier des Plantes 1976; Lavit et al. 1994) to determine a consensus. STATIS is an exploratory data analysis method that is used to obtain an integrated analysis when the data are decomposed into multiple blocks. A consensus matrix is obtained as the matrix with the highest average similarity between all the blocks, in the sense of the Fröbenius norm. This approach can easily be extended to problems where the blocks correspond to different kernels.

More precisely, a similarity measure between kernels is obtained by calculating their cosine in the sense of the Fröbenius norm:

$$\forall m, m' = 1, \ldots, M,$$

$$C_{mm'} = \frac{\langle \mathbf{K}^m, \mathbf{K}^{m'} \rangle_F}{\|\mathbf{K}^m\|_F \|\mathbf{K}^{m'}\|_F} = \frac{\text{Trace}(\mathbf{K}^m \mathbf{K}^{m'})}{\sqrt{\text{Trace}((\mathbf{K}^m)^2)\text{Trace}((\mathbf{K}^{m'})^2)}} \qquad [6.5]$$

$C_{mm'}$ is an extension of the RV coefficient (Robert and Escoufier 1976) to the kernel framework and can be used for exploratory analysis of relationships between kernels (to identify atypical kernels or groups of redundant kernels, respectively, with $C_{mm'}$ values close to 0 and 1).

Therefore, the matrix $\mathbf{C} = (C_{mm'})_{m,m'=1,\ldots,M}$ contains the information about the similarities between kernels and is used to determine a consensus kernel, $\mathbf{K}^*$, which maximizes the average similarity with the other kernels:

$$\max_{\mathbf{v}} \sum_{m=1}^{M} \left\langle \mathbf{K}^{\mathbf{v}} \frac{\mathbf{K}^m}{\|\mathbf{K}^m\|_F} \right\rangle_F \qquad [6.6]$$

with:

$$\sum_{m=1}^{M} \left\langle \mathbf{K}^{\mathbf{v}} \frac{\mathbf{K}^m}{\|\mathbf{K}^m\|_F} \right\rangle_F = \mathbf{v}^\top \mathbf{C} \mathbf{v} \qquad [6.7]$$

for:

$$\mathbf{K}^{\mathbf{v}} = \sum_{m=1}^{M} v_m \mathbf{K}^m$$

and:

$$\mathbf{v} \in \mathbb{R}^M \text{ such that } \|\mathbf{v}\|^2_{\mathbb{R}^M} = 1$$

The solution to the optimization problem of equation [6.6] is given by the spectral decomposition of $\mathbf{C}$. More precisely, if $\mathbf{v} = (v_m)_{m=1,\ldots,M}$ is the first eigenvector of norm 1 for this decomposition, then its entries are all positive (because the matrices $\mathbf{K}^m$ are positive definite) and it maximizes the quantity $\mathbf{v}^\top \mathbf{C} \mathbf{v}$. By defining $\gamma = \frac{\mathbf{v}}{\sum_{m=1}^M v_m}$, a solution satisfying the constraints of equation [6.4] is obtained, which corresponds to a consensual summary of the initial $M$ kernels.

Finally, it should be noted that this method is equivalent to performing a multiple correlation canonical analysis (CCA) between the $M$ feature spaces induced by the $M$ kernels, as proposed in Wang et al. (2008), and in Ren et al. (2013) for a supervised setting and a multiple kernel PCA. Nonetheless, in our approach, only the first axis of the CCA is retained and the constraint on the $\ell_2$ norm ($\|.\|_{\mathbb{R}^M}$) is used to obtain a solution using a simple spectral decomposition. This solution is well suited to problems where the number of kernels is small.

### 6.4.4. *A parsimonious kernel that preserves the topology of the initial data*

The previous proposal aims to obtain consensus information. It therefore tends to give more weights to redundant kernels and disregard the information provided by atypical kernels. However, it may be useful to have a reverse strategy, which gives preferential weighting to additional information rather than redundant information. In our second proposal, *sparse-UMKL*, we establish a criterion that enables the local data geometry to be preserved, in order to more fairly weight the various views carried by the different kernels.

To measure the geometry of the initial space of each kernel, and contrary to Zhuang et al. (2011), we only utilize the information provided by the kernels which we simplify as an undirected $k$-nearest neighbor graph (for a given value $k \in \mathbb{N}^*$), in the sense of the metric induced by $K^m$. This graph is denoted $\mathcal{G}^m$. The adjacency matrix of a combined graph ($\mathcal{G}$) is then obtained by summing the adjacency matrices of the graphs $\mathcal{G}^m$. This matrix ($\mathbf{W}$), of dimension $n \times n$, counts for each pair of observations $i$ and $i'$, the number of times $i'$ is in the $k$ nearest neighbors of $i$ (and vice versa) for one of the $M$ kernels.

A criterion is finally defined, making it possible to find weights $\gamma_m$ that best reproduce the topology represented by $\mathbf{W}$. For this purpose, if $\phi^\gamma$ is the feature map associated with the kernel $K^\gamma = \sum_{m=1}^{M} \gamma_m K^m$, we make sure that when $W_{ii'}$ is large, $\phi^\gamma(x_i)$ and $\phi^\gamma(x_{i'})$ are close in the feature space (and reciprocally). A natural criterion would be to find $\gamma \in \mathbb{R}^M$, such that $\gamma_m \geq 0$ and $\sum_{m=1}^{M} \gamma_m = 1$, minimizing $\sum_{ii'} W_{ii'} \|\phi^\gamma(x_i) - \phi^\gamma(x_{i'})\|_{\mathcal{H}^\gamma}^2$. However, similarly to what is observed by Speicher and Pfeifer (2017) for PCA, this criterion is equivalent to minimizing:

$$\sum_{m=1}^{M} \gamma_m \underbrace{\sum_{i,i'} W_{ii'}(k_{ii}^m + k_{i'i'}^m - 2k_{ii'}^m)}_{=a_m}$$

which has a trivial solution:

$$\gamma_m = \begin{cases} 1 \text{ for } m^* := \arg\min_m a_m \\ 0 \text{ otherwise} \end{cases}$$

Following the idea by Lin et al. (2010), we propose to merely consider a representation of $\phi^\gamma(x_i)$ that corresponds to the similarity of $\phi^\gamma(x_i)$ with the set of the $(\phi^\gamma(x_{i'}))_{i'=1,\ldots,n}$:

$$C_i(\gamma) = \left\langle \phi^\gamma(x_i), \begin{pmatrix} \phi^\gamma(x_1) \\ \vdots \\ \phi^\gamma(x_n) \end{pmatrix} \right\rangle_{\mathcal{H}^\gamma} = \begin{pmatrix} K^\gamma(x_i, x_1) \\ \vdots \\ K^\gamma(x_i, x_n) \end{pmatrix}$$

The following optimization problem is then solved:

$$\min_\gamma \sum_{i,i'=1}^{n} W_{ii} \|C_i(\gamma) - C_{i'}(\gamma)\|_{\mathbb{R}^n}^2$$

for:

$$\gamma \in \mathbb{R}^M \text{ such that } \gamma_m \geq 0 \text{ and } \sum_{m=1}^{M} \gamma_m = 1$$

which is equivalent to:

$$\min_\gamma \sum_{m,m'=1}^{M} \gamma_m \gamma_{m'} S_{mm'} \qquad [6.8]$$

for:

$$\gamma \in \mathbb{R}^M \text{ such that } \gamma_m \geq 0 \text{ and } \sum_{m=1}^M \gamma_m = 1$$

with:

$$S_{mm'} = \sum_{i,i'=1}^N W_{ii'} \langle C_i^m - C_{i'}^m, C_i^{m'} - C_{i'}^{m'} \rangle$$

The matrix $\mathbf{S} = (S_{mm'})_{m,m'=1,\ldots,M}$ is positive and the optimization problem is therefore a standard quadratic programming (QP) problem with linear constraints, which can be solved with the R package quadprog. Moreover, since $\gamma_m \geq 0$, the constraint $\sum_{m=1}^M \gamma_m = 1$ is an $\ell_1$-norm constraint in a QP problem and thus performs a parsimonious selection of kernels (the solution produced will tend to retain only certain non-zero coefficients). This property, which can be seen as a limitation if it is desirable to use all the available kernels, can be overturned by modifying the criterion of equation [6.8], as described in section 6.4.5.

### 6.4.5. *A complete kernel preserving the topology of the initial data*

A simple approach (which we call *full*-**UMKL**) to overcome the parsimony property of the solution to equation [6.8] consists of replacing the constraint on the $\ell_1$ norm by a constraint on the $\ell_2$ norm, similar to what is done in equation [6.6] of STATIS-UMKL:

$$\min_{\mathbf{v}} \sum_{m,m'=1}^M v_m v_{m'} S_{mm'} \qquad [6.9]$$

$$\mathbf{v} \in \mathbb{R}^M \text{ such that } v_m \geq 0 \text{ and } \|\mathbf{v}\|_{\mathbb{R}^M}^2 = 1$$

We then define $\gamma = \frac{\mathbf{v}}{\sum_m v_m}$. This problem is a quadratically constrained quadratic program (QCQP), which is known to be more difficult than linear constrained problems. We propose to use an ADMM-based solution (*Alterning Direction Method of Multipliers*; see Boyd et al. (2011)), which involves rewriting the original problem as:

$$\min_{\mathbf{x} \text{ and } \mathbf{z}} \mathbf{x}^T \mathbf{S} \mathbf{x} + \mathbb{I}_{\{\mathbf{x} \geq 0\}}(\mathbf{x}) + \mathbb{I}_{\{\mathbf{z} \geq 1\}}$$

such that:

$$\mathbf{x} - \mathbf{z} = 0$$

and where the final solution can be obtained by $\gamma := \frac{\mathbf{z}}{\sum_m z_m}$.

## 6.5. Application

The objective of this chapter is to describe the different steps to integrate and analyze the data from Sunagawa et al. (2015) using unsupervised kernel methods available in the R package mixKernel (version 0.3). The session information corresponding to the results presented in this section is available in section 6.6.

The data used were collected as part of the Tara Ocean expedition (Karsenti et al. 2011; Bork et al. 2015) that facilitated the study of planktonic communities by making a set of ocean metagenomics data coupled with environmental measurements available. The analysis focuses on the study of 139 samples collected from 68 sampling stations and three ocean depths: the surface (SRF), the maximum chlorophyll layer (DCM) and the mesopelagic zone (MES). The 68 selected stations are located on eight different oceans and seas: the Indian Ocean (IO), the Mediterranean Sea (MS), the North Atlantic Ocean (NAO), the North Pacific Ocean (NPO), the Red Sea (RS), the South Atlantic Ocean (SAO), the South Pacific Ocean (SPO) and the Austral Ocean (SO).

To simplify the analysis and obtain reasonable computation times for illustration purposes, we will use only a subset of the data analyzed in Mariette and Villa-Vialaneix (2018). This subset, available in the mixKernel package, includes 1% of the 35,650 operational taxonomic units (OTUs) in prokaryotes and 39,246 bacteria genes. This subset was selected randomly.

The mixKernel package is installed and loaded under R with the commands:

```
install.package(mixKernel)
library(mixKernel)
```

### 6.5.1. *Loading Tara Ocean data*

The datasets, previously normalized, are provided in the form of matrices whose row names, representing the different samples, are the same:

```
data(TARAoceans)
# more details with: ?TARAoceans
# we check the dimension of the data:
lapply(list("phychem" = TARAoceans$phychem,
            "pro.phylo" = TARAoceans$pro.phylo,
            "pro.NOGs" = TARAoceans$pro.NOGs),
       dim)
## $phychem
## [1] 139  22
##
## $pro.phylo
## [1] 139 356
##
## $pro.NOGs
## [1] 139 638
```

### 6.5.2. *Data integration by kernel approaches*

For each dataset, a kernel is computed using the function `compute.kernel`, which offers the possibility to choose among the linear, Gaussian, Poisson (Witten 2011), phylogenetic (Lozupone and Knight 2005; Lozupone et al. 2007) or abundance kernels (Sørensen 1948; Bray and Curtis 1957). Users also have the possibility to define their own kernel with the parameter `kernel.func` (for more information: ?`compute.kernel`).

The results are returned as a list in which the element `kernel` stores the kernel matrix. The resulting kernels are symmetrical matrices whose dimension is equal to the number of samples, that is, the number of rows existing in the original dataset.

```
phychem.kernel <- compute.kernel(TARAoceans$phychem,
                                 kernel.func = "linear")
pro.phylo.kernel <- compute.kernel(TARAoceans$pro.phylo,
                                   kernel.func = "abundance")
pro.NOGs.kernel <- compute.kernel(TARAoceans$pro.NOGs,
                                  kernel.func = "abundance")

# check dimensions
dim(pro.NOGs.kernel$kernel)

## [1] 139 139
```

Once the kernels are computed, the function `cim.kernel` provides a global view of the structure of their correlations:

```
cim.kernel(phychem = phychem.kernel,
           pro.phylo = pro.phylo.kernel,
           pro.NOGs = pro.NOGs.kernel,
           method = "square")
```

Figure 6.3 shows that pro.phylo and pro.NOGs are the most correlated pair of kernels. This result is expected because these kernels both provide a view of the same community: the bacterial community.

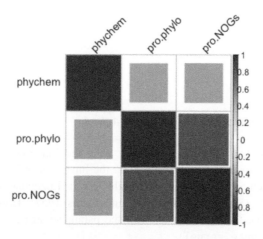

**Figure 6.3.** *Similarities between kernels computed using the STATIS-UMKL (the color and the area of the square represent the level of similarity between two kernels)*

The function combine.kernels offers three different methods for combining several kernels: STATIS-UMKL, *sparse-UMKL* and *full-UMKL*, which are all described in sections 6.4.3, 6.4.4 and 6.4.5. This function returns a meta-kernel that can be used as an input to the function kernel.pca. The three methods provide complementary information and the choice will have to be made on the basis of the research question that was posed. The STATIS-UMKL approach yields an overview of the common information between the different datasets. The *full-UMKL* method returns a meta-kernel that minimizes the distortion between the different kernels provided as inputs, and *sparse-UMKL* is a parsimonious version of *full-UMKL*, which selects the most relevant kernels.

```
meta.kernel <- combine.kernels(phychem = phychem.kernel,
                               pro.phylo = pro.phylo.kernel,
                               pro.NOGs = pro.NOGs.kernel,
                               method = "full-UMKL")
```

### 6.5.3. *Exploratory analysis: kernel PCA*

A kernel PCA can then be obtained from the combined kernel using the function kernel.pca. The parameter ncomp allows the choice of the number of components to be extracted from the kernel PCA.

```
kpc.res <- kernel.pca(meta.kernel, ncomp = 10)
```

The projection of the individuals onto the axes of the kernel PCA can be displayed using the function plotIndiv (the result in Figure 6.4A):

```
all.depths <- levels(factor(TARAoceans$sample$depth))
depth.pch <- c(20, 17, 4, 3)[match(TARAoceans$sample$depth,
                    all.depths)]
plotIndiv(kpc.res,
          comp = c(1, 2),
          ind.names = FALSE,
          legend = TRUE,
          group = as.vector(TARAoceans$sample$ocean),
          col.per.group = c("#f99943", "#44a7c4", "#05b052",
                            "#2f6395",
                            "#bb5352", "#87c242", "#07080a",
                            "#92bbdb"),
          pch = depth.pch,
          pch.levels = TARAoceans$sample$depth,
          legend.title = "Ocean / Sea",
          title = "Projection of TARA Oceans stations",
          size.title = 10,
          legend.title.pch = "Depth")
```

These results are similar to those presented in Sunagawa et al. (2015): the samples are separated by their depths, that is, SRF, DCM or MES, with strong atypicity in MES samples. The variance explained by each kernel PCA axis can be obtained with the function plot (Figure 6.4) and assist the user in choosing the number of components to extract. Here, the first axis explains 20% of the total variance.

```
plot(kpc.res)
```

In the remainder of this chapter, we focus on the information carried by the first component only. In order to study the influence of the variables of the different datasets, their values are randomly permuted using the function permute.kernel.pca. Figure 6.5 presents the result of the permutation of the physicochemical variables at the level of the variables themselves (kernel phychem), of the abundances in OTU of the kernel pro.phylo at the phylum level (OTU phyla are stored in the second column, named Phylum, of the taxonomic annotation available through the entry taxonomy of the object TARAoceans) and gene abundances of the kernel pro.NOGs at the GO level (GOs are available from the dataset entry GO):

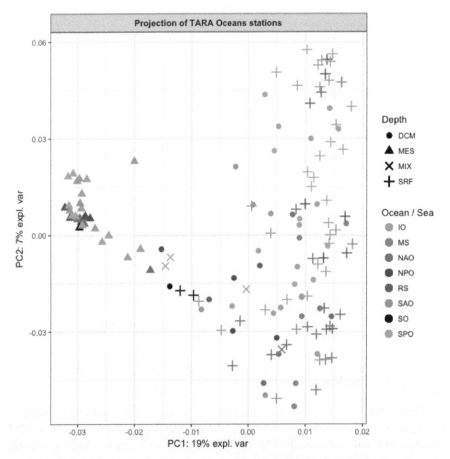

**Figure 6.4A.** *Projection of individuals onto the first two axes of the kernel PCA (the colors and shapes, respectively, represent the ocean regions and depths)*

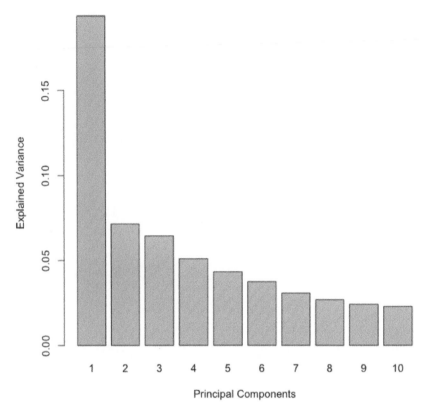

**Figure 6.4B.** *Variance explained by the first 10 axes of the kernel PCA performed on the kernel obtained by the full-UMKL approach*

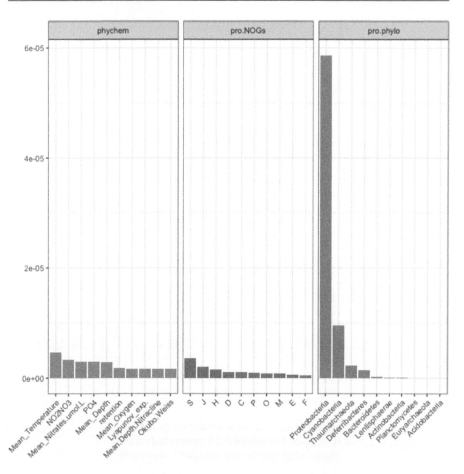

**Figure 6.5.** *The five most important variables for each of the three datasets, ordered by decreasing importance*

```
head(TARAoceans$taxonomy[ ,"Phylum"], 10)

## [1] Actinobacteria    Proteobacteria    Proteobacteria
## [4] Gemmatimonadetes  Actinobacteria    Actinobacteria
## [7] Proteobacteria    Proteobacteria    Proteobacteria
## ...
## 56 Levels: Acidobacteria Actinobacteria aquifer1 ...
#    WCHB1-60

head(TARAoceans$GO, 10)

## [1] NA   NA   "K"  NA   NA   "S"  "S"  "S"  NA   "S"

# here we set a seed for reproducible results with this
#       tutorial
set.seed(17051753)
kpc.res <- kernel.pca.permute(kpc.res, ncomp = 1,
            phychem = colnames(TARAoceans$phychem),
            pro.phylo = TARAoceans$taxonomy[ ,"Phylum"],
            pro.NOGs = TARAoceans$GO)
```

The results, presented in Figure 6.5, can be visualized with the function plotVar.kernel.pca. The parameter ndisplay enables the definition of the number of variables to display for each kernel:

```
plotVar.kernel.pca(kpc.res, ndisplay = 10, ncol = 3)
```

*Proteobacteria* is the most important variable in the kernel, pro.phylo. The relative abundance values of *Proteobacteria*, are displayed as colors on the left-hand side of Figure 6.6:

```
selected <- which(TARAoceans$taxonomy[ ,"Phylum"] ==
                "Proteobacteria")
proteobac.sample <- apply(TARAoceans$pro.phylo[ ,selected],
                1, sum)
proteobac.sample <- proteobac.sample /
                apply(TARAoceans$pro.phylo, 1, sum)
colfunc <- colorRampPalette(c("royalblue", "red"))
col.proteo <- colfunc(length(proteobac.sample))
col.proteo <- col.proteo[rank(proteobac.sample,
                ties = "first")]
plotIndiv(kpc.res,
        comp = c(1, 2),
        ind.names = FALSE,
```

```
             legend = FALSE,
             group = c(1:139),
             col = col.proteo,
             pch = depth.pch,
             pch.levels = TARAoceans$sample$depth,
             legend.title = "Ocean / Sea",
             title = "Representation of Proteobacteria
                     abundance",
             legend.title.pch = "Depth")
```

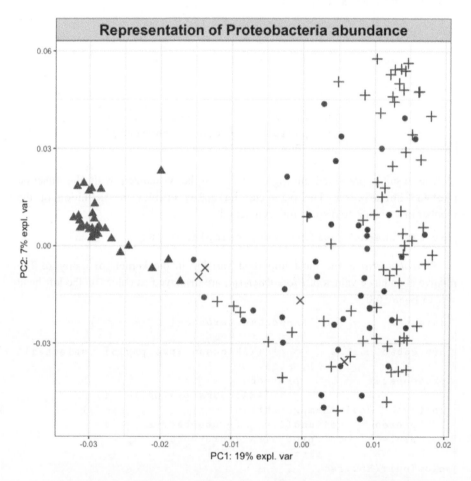

**Figure 6.6A.** *Projection of individuals on the first two axes of the kernel PCA: the colors represent the relative abundance in* Proteobacteria

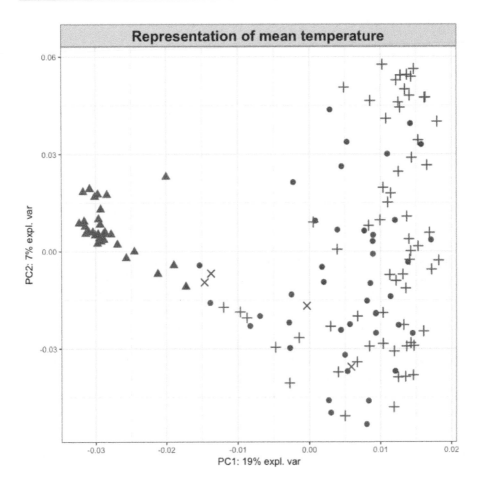

**Figure 6.6B.** *Sample projection onto the first two axes of the kernel PCA: the colors represent the temperature (blue for cold waters and red for warm waters)*

Similarly, the temperature is the most important variable of the phychem kernel. The temperatures can then be displayed on the figure representing the projection of the kernel PCA (on the right-hand side of Figure 6.6):

```
col.temp <- colfunc(length(TARAoceans$phychem[ ,4]))
col.temp <- col.temp[rank(TARAoceans$phychem[ ,4],
                    ties = "first")]
plotIndiv(kpc.res,
          comp = c(1, 2),
```

```
              ind.names = FALSE,
              legend = FALSE,
              group = c(1:139),
              col = col.temp,
              pch = depth.pch,
              pch.levels = TARAoceans$sample$depth,
              legend.title = "Ocean / Sea",
              title = "Representation of mean temperature",
              legend.title.pch = "Depth")
```

For the two graphs in Figure 6.6, the color gradients observed on the first axis of the kernel PCA between the left-hand side (low *Proteobacteria* abundances and low temperatures) and the right side (high abundances of *Proteobacteria* and high temperatures) confirm the contribution of the *Proteobacteria* and temperature to the definition of the first axis. These variables specifically structure the data subsample (which was already observed in the preliminary study by Mariette and Villa-Vialaneix (2018) and is consistent with the results discussed in (Sunagawa et al. 2015)). These variables separate the samples according to their depth, particularly those from the mesopelagic zone of the other samples (the deepest and coldest, triangle-shaped in the figure). The second axis shows an inverse association between samples at high temperature (top), which correspond to samples in which the abundance of *Proteobacteria* is more moderate. In an integrated manner, the analysis allowed the extraction of the most structuring variables of variability between samples and a representation that yields the association between these variables and the characteristics of the samples.

## 6.6. Session information for the results of the example

```
sessionInfo()

## R version 3.4.3 (2017-11-30)
## Platform: x86_64-pc-linux-gnu (64-bit)
## Running under: Ubuntu 16.04.5 LTS
##
## Matrix products: default
## BLAS: /usr/local/lib/R/lib/libRblas.so
## LAPACK: /usr/local/lib/R/lib/libRlapack.so
##
## locale:
##  [1] LC_CTYPE=fr_FR.UTF-8       LC_NUMERIC=C
##  [3] LC_TIME=fr_FR.UTF-8        LC_COLLATE=fr_FR.UTF-8
##  [5] LC_MONETARY=fr_FR.UTF-8    LC_MESSAGES=fr_FR.UTF-8
##  [7] LC_PAPER=fr_FR.UTF-8       LC_NAME=C
```

```
## [9] LC_ADDRESS=C              LC_TELEPHONE=C
## [11] LC_MEASUREMENT=fr_FR.UTF-8 LC_IDENTIFICATION=C
##
## attached base packages:
## [1] stats    graphics  grDevices utils    datasets  methods
#         base
##
## other attached packages:
## [1] mixKernel_0.3   mixOmics_6.3.2   ggplot2_2.2.1
#            lattice_0.20-35
## [5] MASS_7.3-50      knitr_1.20
##
## loaded via a namespace (and not attached):
##  [1] Biobase_2.38.0     tidyr_0.8.1        splines_3.4.3
##  [4] jsonlite_1.5       foreach_1.4.4      ellipse_0.4.1
##  [7] shiny_1.1.0        assertthat_0.2.0   stats4_3.4.3
## [10] phyloseq_1.22.3    yaml_2.1.19        corrplot_0.84
## [13] pillar_1.2.3       backports_1.1.2    quadprog_1.5-5
## [16] glue_1.2.0         digest_0.6.15
#            manipulateWidget_0.9.0
## [19] RColorBrewer_1.1-2 promises_1.0.1     XVector_0.18.0
## [22] colorspace_1.3-2   psych_1.8.4        htmltools_0.3.6
## [25] httpuv_1.4.3       Matrix_1.2-14      plyr_1.8.4
## [28] pkgconfig_2.0.2    zlibbioc_1.24.0    purrr_0.2.5
## [31] xtable_1.8-2       corpcor_1.6.9      scales_0.5.0
## [34] RSpectra_0.13-1    later_0.7.2        tibble_1.4.2
## [37] mgcv_1.8-23        IRanges_2.12.0     BiocGenerics_0.24.0
## [40] lazyeval_0.2.1     mnormt_1.5-5       survival_2.42-3
## [43] magrittr_1.5       mime_0.5           evaluate_0.10.1
## [46] nlme_3.1-137       foreign_0.8-70     vegan_2.5-2
## [49] data.table_1.11.4  tools_3.4.3        matrixStats_0.53.1
## [52] stringr_1.3.1      S4Vectors_0.16.0   munsell_0.4.3
## [55] cluster_2.0.7-1    bindrcpp_0.2.2     Biostrings_2.46.0
## [58] ade4_1.7-11        compiler_3.4.3     rlang_0.2.1
## [61] rhdf5_2.22.0       grid_3.4.3         iterators_1.0.9
## [64] biomformat_1.6.0   htmlwidgets_1.2    crosstalk_1.0.0
## [67] igraph_1.2.2       miniUI_0.1.1.1     labeling_0.3
## [70] rmarkdown_1.9      multtest_2.34.0    gtable_0.2.0
## [73] codetools_0.2-15   rARPACK_0.11-0     reshape2_1.4.3
## [76] R6_2.2.2           gridExtra_2.3      dplyr_0.7.6
## [79] bindr_0.1.1        rprojroot_1.3-2    LDRTools_0.2-1
## [82] permute_0.9-4      ape_5.2            stringi_1.2.2
## [85] parallel_3.4.3     Rcpp_1.0.0         rgl_0.99.16
## [88] tidyselect_0.2.4
```

## 6.7. References

Allen, G.I. (2013). Automatic feature selection via weighted kernels and regularization. *Journal of Computational and Graphical Statistics*, 22(2), 284–299.

Ambroise, C. and Govaert, G. (1996). Analyzing dissimilarity matrices via Kohonen maps. In *Proceedings of 5th Conference of the International Federation of Classification Societies (IFCS 1996)*, Springer, New York.

Ambroise, C., Dehman, A., Neuvial, P., Rigaill, G., Vialaneix, N. (2019). Adjacency-constrained hierarchical clustering of a band similarity matrix with application to genomics. *Algorithms for Molecular Biology*, 14, 22.

Andras, P. (2002). Kernel-Kohonen networks. *International Journal of Neural Systems*, 12, 117–135.

Aronszajn, N. (1950). Theory of reproducing kernels. *Transactions of the American Mathematical Society*, 68(3), 337–404.

Ben-Hur, A., Ong, C.S., Sonnenburg, S., Schölkopf, B., Rätsch, G. (2008). Support vector machines and kernels for computational biology. *PLoS Computational Biology*, 4(10), e1000173.

Bersanelli, M., Mosca, E., Remondini, D., Giampieri, E., Sala, C., Castellani, G., Milanesi, L. (2016). Methods for the integration of multi-omics data: Mathematical aspects. *BMC Bioinformatics*, 17(Suppl. 2), S15.

Borgwardt, K.M., Ong, C.S., Schönauer, S., Vishwanathan, S., Smola, A.J., Kriegel, H.-P. (2005). Protein function prediction via graph kernels. *Bioinformatics*, 21, i47–i56.

Bork, P., Bowler, C., de Vargas, C., Gorsky, G., Karsenti, E., Wincker, P. (2015). Tara oceans studies plankton at planetary scale. *Science*, 348(6237), 873–873.

Boyd, S., Parikh, N., Chu, E., Peleato, B., Eckstein, J. (2011). Distributed optimization and statistical learning via the alterning direction method of multipliers. *Foundations and Trends in Machine Learning*, 3(1), 1–122.

Bray, R.J. and Curtis, J. (1957). An ordination of the upland forest communities of Southern Wisconsin. *Ecological Monographs*, 27(4), 325–349.

Brouard, C., Shen, H., Dürkop, K., d'Alché Buc, F., Böcker, S., Rousu, J. (2016). Fast metabolite identification with input output kernel regression. *Bioinformatics*, 32(12), i28–i36.

Chen, Y., Garcia, E., Gupta, M., Rahimi, A., Cazzanti, L. (2009). Similarity-based classification: Concepts and algorithm. *Journal of Machine Learning Research*, 10, 747–776.

Collins, M., Dasgupta, S., Schapire, R.E. (2001). A generalization of principal component analysis to the exponential family. In *Advances in Neural Information Processing Systems (NIPS)*, vol. 13, Dietterich, T., Becker, S., Ghahramani, Z. (eds). MIT Press, Cambridge, MA.

Cottrell, M., Olteanu, M., Rossi, F., Villa-Vialaneix, N. (2016). Theoretical and applied aspects of the self-organizing maps. In *Advances in Self-Organizing Maps and Learning Vector Quantization. Advances in Intelligent Systems and Computing*, vol. 428, Merényi, E., Mendenhall, M., O'Driscoll, P. (eds). Springer, Cham.

Dührkop, K., Shen, H., Meusel, M., Rousu, J., Böcker, S. (2015). Searching molecular structure databases with tandem mass spectra using CSI:FingerID. *Proceedings of the National Academy of Sciences of the United States of America*, 112(41), 12580–12585.

El Golli, A., Rossi, F., Conan-Guez, B., Lechevallier, Y. (2006). Une adaptation des cartes auto-organisatrices pour des données décrites par un tableau de dissimilarités. *Revue de statistique appliquée*, LIV(3), 33–64.

Franzosa, E.A., Hsu, T., Sirota-madi, A., Shafquat, A., Abu-Ali, G., Morgan, X.C., Huttenhower, C. (2015). Sequencing and beyond: Integrating molecular "omics" for microbial community profiling. *Nature Reviews Microbiology*, 13(6), 360–372.

Gärtner, T., Flach, P., Wrobel, S. (2003). On graph kernels: Hardness results and efficient alternatives. In *Proceedings of the Annual Conference on Computational Learning Theory*. Lecture Notes in Computer Science, vol. 2777. Springer, Berlin, Heidelberg.

Goldfarb, L. (1984). A unified approach to pattern recognition. *Pattern Recognition*, 17(5), 575–582.

Gönen, M. and Alpaydin, E. (2011). Multiple kernel learning algorithms. *Journal of Machine Learning Research*, 12, 2211–2268.

Gönen, M. and Margolin, A.A. (2014). Localized data fusion for kernel k-means clustering with application to cancer biology. In *Proceedings of Advances in Neural Information Processing Systems*, vol. 27, Ghahramani, Z., Welling, M., Cortes, C., Lawrence, N., Weinberger, K. (eds). Neural Information Processing Systems Foundation, Inc. (NeurIPS).

Graepel, T., Burger, M., Obermayer, K. (1998). Self-organizing maps: Generalizations and new optimization techniques. *Neurocomputing*, 21, 173–190.

Hammer, B. and Hasenfuss, A. (2010). Topographic mapping of large dissimilarity data sets. *Neural Computation*, 22(9), 2229–2284.

Haussler, D. (1999). Convolution kernels on discrete structures. Technical Report, UCS-CRL-99-10.

He, X. and Niyogi, P. (2003). Locality preserving projections. *Proceedings of Advances in Neural Information Processing Systems*, 4.

Hofmann, D., Gisbrecht, A., Hammer, B. (2015). Efficient approximations of robust soft learning vector quantization for non-vectorial data. *Neurocomputing*, 147, 96–106.

Huang, H.-C., Chuang, Y.-Y., Chen, C.-S. (2012). Multiple kernel fuzzy clustering. *IEEE Transactions on Fuzzy Systems*, 20(1), 120–134.

Jaccard, P. (1912). The distribution of the flora in the alpine zone. *New Phytologist*, 11(2), 37–50.

Karsenti, E., Acinas, S., Bork, P., Bowler, C., de Vargas, C., Raes, J., Sullivan, M., Arendt, D., Benzoni, F., Claverie, J. et al. (2011). A holistic approach to marine eco-systems biology. *PLoS Biology*, 9(10), e1001177.

Kaufman, L. and Rousseeuw, P. (1987). Clustering by means of medoids. In *Statistical Data Analysis Based on the L1-Norm and Related Methods*, Dodge, Y. (ed.). Birkhäuser, Basel.

Kohonen, T. (2001). *Self-Organizing Maps*, 3rd edition. Springer, Berlin/Heidelberg/New York.

Kohohen, T. and Somervuo, P. (1998). Self-organizing maps of symbol strings. *Neurocomputing*, 21, 19–30.

Kohonen, T. and Somervuo, P. (2002). How to make large self-organizing maps for nonvectorial data. *Neural Networks*, 15(8), 945–952.

Kondor, R. and Lafferty, J. (2002). Diffusion kernels on graphs and other discrete structures. In *Proceedings of the 19th International Conference on Machine Learning*, Sammut, C. and Hoffmann, A. (eds). Morgan Kaufmann Publishers Inc., San Francisco, CA/Sydney.

Kristensen, V.N., Lingjærde, O.C., Russnes, H.G., Vollan, H.K.M., Frigessi, A., Børresen-Dale, A.-L. (2014). Principles and methods of integrative genomic analyses in cancer. *Nature Reviews Cancer*, 14(5), 299–313.

Kruskal J.B. (1964). Multidimensional scaling by optimizing goodness of fit to a nonmetric hypothesis. *Psychometrika*, 29(1), 1–27.

Kumar, S., Mohri, M., Talwalkar, A. (2012). Sampling techniques for the Nyström method. *Journal of Machine Learning Research*, 13, 981–1006.

Lanckriet, G., Cristianini, N., Bartlett, P., El Ghaoui, L., Jordan, M. (2004). Learning the kernel matrix with semidefinite programming. *Journal of Machine Learning Research*, 5, 27–72.

Lavit, C., Escoufier, Y., Sabatier, R., Traissac, P. (1994). The ACT (STATIS method). *Computational Statistics and Data Analysis*, 18(1), 97–119.

Lee, J. and Verleysen, M. (2007). *Nonlinear Dimensionality Reduction*. Springer, New York/London.

L'Hermier des Plantes, H. (1976). Structuration des tableaux à trois indices de la statistique. PhD Thesis, Université de Montpellier.

Lin, Y., Liu, T., Fuh, C.-S. (2010). Multiple kernel learning for dimensionality reduction. *IEEE Transactions on Pattern Analysis and Machine Intelligence*, 33, 1147–1160.

Lozupone, C.A. and Knight, R. (2005). UniFrac: A new phylogenetic method for comparing microbial communities. *Applied and Environmental Microbiology*, 71(12), 8228–8235.

Lozupone, C.A., Hamady, M., Kelley, S.T., Knight, R. (2007). Quantitative and qualitative $\beta$ diversity measures lead to different insights into factors that structure microbial communities. *Applied and Environmental Microbiology*, 73(5), 1576–1585.

MacDonald, D. and Fyfe, C. (2000). The kernel self organising map. *Proceedings of 4th International Conference on Knowledge-based Intelligence Engineering Systems and Applied Technologies.*

Mahé, P. and Vert, J. (2009). Graph kernels based on tree patterns for molecules. *Machine Learning*, 75, 3–35.

Mariette, J. and Villa-Vialaneix, N. (2018). Unsupervised multiple kernel learning for heterogeneous data integration. *Bioinformatics*, 34(6), 1009–1015.

Mariette, J., Olteanu, M., Boelaert, J., Villa-Vialaneix, N. (2014). Bagged kernel SOM. In *Advances in Self-Organizing Maps and Learning Vector Quantization. Advances in Intelligent Systems and Computing*, vol. 295, Villmann, T., Schleif, F., Kaden, M., Lange, M. (eds). Springer, Cham.

Mariette, J., Olteanu, M., Villa-Vialaneix, N. (2017). Efficient interpretable variants of online SOM for large dissimilarity data. *Neurocomputing*, 225, 31–48.

Meher, P.K., Sahu, T.K., Rao, A., Wahi, S. (2016). Identification of donor splice sites using support vector machine: A computational approach based on positional, compositional and dependency features. *Algorithms for Molecular Biology*, 11(16), 16.

Noble, W.S. (2004). Support vector machine applications in computational biology. In *Kernel Methods in Computational Biology*. MIT Press, Cambridge, MA/London.

Olteanu, M. and Villa-Vialaneix, N. (2015). On-line relational and multiple relational SOM. *Neurocomputing*, 147, 15–30.

Qin, J., Lewis, D.P., Noble, W.S. (2003). Kernel hierarchical gene clustering from microarray expression data. *Bioinformatics*, 19(16), 2097–2104.

Qiu, J., Hue, M., Ben-Hur, A., Vert, J.-P., Stafford, N.W. (2007). A structural alignment kernel for protein structures. *Bioinformatics*, 23(9), 1090–1098.

Ramon, J. and Gärtner, T. (2003). Expressivity versus efficiency of graph kernels. In *Proceedings of First International Workshop on Mining Graphs, Trees and Sequences (Held with ECML/PKDD'03)*, Washio, T. and de Raedt, L. (eds). Cavtat-Dubrovnik.

Rapaport, F., Zinovyev, A., Dutreix, M., Barillot, E., Vert, J. (2007). Classification of microarray data using gene networks. *BMC Bioinformatics*, 8, 35.

Rappoport, N. and Shamir, R. (2018). Multi-omic and multi-view clustering algorithms: Review and cancer benchmark. *Nucleic Acids Research*, 46(20), 10546–10562.

Ren, S., Ling, P., Yang, M., Ni, Y., Zong, Z. (2013). Multi-kernel PCA with discriminant manifold for hoist monitoring. *Journal of Applied Sciences*, 13(20), 4195–4200.

Ritchie, M.D., Holzinger, E.R., Li, R., Pendergrass, S.A., Kim, D. (2015). Methods of integrating data to uncover genotype-phenotype interactions. *Nature Reviews Genetics*, 16(2), 85–97.

Robert, P. and Escoufier, Y. (1976). A unifying tool for linear multivariate statistical methods: The RV-coefficient. *Applied Statistics*, 25(3), 257–265.

Rossi, F. (2014). How many dissimilarity/kernel self organizing map variants do we need? In *Advances in Self-organizing Maps and Learning Vector Quantization. Advances in Intelligent Systems and Computing*, Villmann, T., Schleif, F., Kaden, M., Lange, M. (eds). Springer, Cham.

Rossi, F., Hasenfuss, A., Hammer, B. (2007). Accelerating relational clustering algorithms with sparse prototype representation. In *Proceedings of the 6th Workshop on Self-organizing Maps (WSOM 07)*. Neuroinformatics Group, Bielefield University.

Schleif, F.-M. and Tino, P. (2015). Indefinite proximity learning: A review. *Neural Computation*, 27(10), 2039–2096.

Schoenberg, I. (1935). Remarks to Maurice Fréchet's article "Sur la définition axiomatique d'une classe d'espace distanciés vectoriellement applicable sur l'espace de Hilbert". *Annals of Mathematics*, 36, 724–732.

Schölkopf, B., Smola, A., Müller, K. (1998). Nonlinear component analysis as a kernel eigenvalue problem. *Neural Computation*, 10(5), 1299–1319.

Schölkopf, B., Tsuda, K., Vert, J. (2004). *Kernel Methods in Computational Biology*. MIT Press, London.

Shen, H., Dührkop, K., Böcher, S., Rousu, J. (2014). Metabolite identification through multiple kernel learning on fragmentation trees. *Bioinformatics*, 30(12), i157–164.

Singh, A., Gautier, B., Shannon, C.P., Rohart, F., Vacher, M., Tebbut, S.J., Lê Cao, K.-A. (2018). DIABLO: From multi-omics assays to biomarker discovery, an integrative approach. *Bioinformatics*, 35(17), 3055–3062.

Smola, A. and Kondor, R. (2003). Kernels and regularization on graphs. In *Proceedings of the Conference on Learning Theory (COLT) and Kernel Workshop. Lecture Notes in Computer Science*, vol. 2777, Warmuth, M. and Schölkopf, B. (eds). Springer, Berlin/Heidelberg/Washington, DC.

Sørensen, T.J. (1948). A method of establishing groups of equal amplitude in plant sociology based on similarity of species and its application to analyses of the vegetation on Danish commons. *Biologiske Skrifter/Kongelige Danske Videnskabernes Selskab*, 5(4), 1–34.

Speicher, N.K. and Pfeifer, N. (2015). Integrating different data types by regularized unsupervised multiple kernel learning with application to cancer subtype discovery. *Bioinformatics*, 31(12), i268–i275.

Speicher, N.K. and Pfeifer, N. (2017). Towards multiple kernel principal component analysis for integrative analysis of tumor samples. *Journal of Integrative Bioinformatics*, 14(2), 20170019.

Sunagawa, S., Coelho, L., Chaffron, S., Kultima, J., Labadie, K., Salazar, F., Djahanschiri, B., Zeller, G., Mende, D., Alberti, A. et al. (2015). Structure and function of the global ocean microbiome. *Science*, 348(6237).

Tang, W., Lu, Z., Dhillon, I. (2009). Clustering with multiple graphs. In *Proceedings of IEEE International Conference on Data Mining (ICDM)*, Wang, W., Kargupta, H., Ranka, S., Yu, P.S., Wu, X. (eds). IEEE Computer Society, Miami, FL.

Tsuda, K. (1999). Support vector classifier with asymmetric kernel functions. In *Proceedings of the 7th European Symposium on Artificial Neural Network (ESANN 1999)*, Verleysen, M. (ed.). D-Facto Public, Bruges.

Vega-Pons, S. and Ruiz-Schulcloper, J. (2011). A survey of clustering ensemble algorithms. *International Journal of Pattern Recognition and Artificial Intelligence*, 25(3), 337–372.

Vert, J.-P. (2002). A tree kernel to analyse phylogenetic profiles. *Bioinformatics*, 18(Suppl. 1), S276–S284.

Vert, J.-P. (2007). Kernel methods in genomics and computational biology. In *Kernel Methods in Bioengineering, Signal and Image Processing*, Camps-Valls, G., Rojo-Alvarez, J.L., Martinez-Ramon, M. (eds). Idea Group, Hershey, PA.

Vert, J.-P. and Kanehisa, M. (2003). Extracting active pathways from gene expression data. *Bioinformatics*, 19(Suppl. 2), ii238–ii244.

Villa, N. and Rossi, F. (2007). A comparison between dissimilarity SOM and kernel SOM for clustering the vertices of a graph. In *6th International Workshop on Self-organizing Maps (WSOM 2007)*. Neuroinformatics Group, Bielefield University.

Vishwanathan, S., Schraudolph, N.N., Kondor, R., Borgwardt, K.M. (2010). Graph kernels. *Journal of Machine Learning Research*, 11, 1201–1242.

Wang, Z., Chen, S., Sun, T. (2008). MultiK-MHKS: A novel multiple kernel learning algorithm. *IEEE Transactions on Pattern Analysis and Machine Intelligence*, 30(2), 348–353.

Wang, B., Zhu, J., Pierson, E., Ramazzotti, D., Batzoglou, S. (2017). Visualization and analysis of single-cell RNA-seq data by kernel-based similarity learning. *Nature Methods*, 14, 414–416.

Ward, J.H. (1963). Hierarchical grouping to optimize an objective function. *Journal of the American Statistical Association*, 53(301), 236–244.

Williams, C. and Seeger, M. (2000). Using the Nyström method to speed up kernel machines. In *Advances in Neural Information Processing Systems (Proceedings of NIPS 2000)*, vol. 13, Leen, T., Dietterich, T., Tresp, V. (eds). Neural Information Processing Systems Foundation, Denver, CO.

Witten, D. (2011). Classification and clustering of sequencing data using a Poisson model. *The Annals of Applied Statistics*, 5(4), 2493–2518.

Young, G. and Householder, A. (1938). Discussion of a set of points in terms of their mutual distances. *Psychometrika*, 3, 19–22.

Yu, S., Tranchevent, L., Liu, X., Glanzel, W., Suykens, J.A., de Moor, B., Moreau, Y. (2012). Optimized data fusion for kernel k-means clustering. *IEEE Transactions on Pattern Analysis and Machine Intelligence*, 34(5), 1031–1039.

Zhao, B., Kwok, J., Zhang, C. (2009). Multiple kernel clustering. In *Proceedings of the 2009 SIAM International Conference on Data Mining (SDM)*, Apte, C., Park, H., Wang, K., Zaki, M. (eds). SIAM, Philadelphia, PA.

Zhuang, J., Wang, J., Hoi, S., Lan, X. (2011). Unsupervised multiple kernel clustering. *Journal of Machine Learning Research: Workshop and Conference Proceedings*, 20, 129–144.

# 7
# Multivariate Models for Data Integration and Biomarker Selection in 'Omics Data

Sébastien DÉJEAN[1] and Kim-Anh LÊ CAO[2]
[1] Institut de Mathématiques de Toulouse, CNRS, Université de Toulouse, France
[2] School of Mathematics and Statistics, Melbourne Integrative Genomics, Australia

## 7.1. Introduction

The advent of high-throughput technologies has enabled new opportunities for biological and medical research discoveries, particularly with the rise of novel 'omics technologies to measure the expression or abundance of different molecular features (transcriptomics, proteomics, metabololomics, etc.) on the same individuals or samples. However, biological systems cannot be understood by the analysis of single-type datasets, as the regulation of the system certainly occurs at many functional levels (Zhang et al. 2009). There is therefore a need to combine these 'omics data to obtain a better picture of a biological system as a whole. In the field of computational biology and bioinformatics, data integration has been recognized as a major challenge for many years (Goble and Stevens 2008; Gomez-Cabrero et al. 2014), and especially for 'omics data. One of the main challenges we face is the

---

For a color version of all figures in this chapter, see: http://www.iste.co.uk/froidevaux/biologicaldata.zip.

heterogeneous nature of the data, which are generated using different technological platforms and are measured on different scales. The aim of 'omics data integration is to not only provide a unified view of the data that come from disparate sources and are generated using various technologies, but also an understanding of a biological system under mathematical models that can mechanistically describe the relationships between their components. Data integration may therefore elucidate signaling pathways, enabling us to gain evolutionary insight and understand the role of changes in transcriptional regulatory networks.

In order to achieve the full potential of 'omics studies, there is a need to develop novel analysis tools that are able to handle the volume of data, make biological sense, improve our understanding of the biological system and provide useful and insightful visualizations. Several approaches have recently been proposed to integrate data. They fall into two main categories: knowledge-driven and data-driven approaches.

Knowledge-driven approaches use public data repositories sometimes combined with in-house data using network modeling approaches (Joyce and Palsson 2006; Gomez-Cabrero et al. 2014). The public repositories include, among others, the Encyclopaedia of DNA Elements Project (The ENCODE (ENCyclopedia Of DNA Elements) Project (The ENCODE Project Consortium 2004)), which aims to characterize a set of animal models, tissue or cell lines given their mRNA expression, histone marks, Chip-Seq and the location of active regulatory regions; the Functional Annotation of the Mammalian Genome (FANTOM (Carninci et al. 2005)), which focuses on characterizing the transcriptional regulatory network; The Cancer Genome Atlas Project (TCGA (McLendon et al. 2008)), which aims to generate insights into the heterogeneity of different cancer subtypes by creating a map of molecular alterations for every type of cancer at multiple levels (mRNA, miRNA, protein, DNA methylation, copy number alterations and somatic chromosomal aberrations). Data-driven approaches are focused on extracting correlations or associations between 'omics datasets as an indicator of a consensus between the functional levels. The 'omics data are often generated on the same samples (patients or biological specimens). Several statistical tools have been proposed to integrate two 'omics datasets, such as unsupervised multivariate exploratory analysis for data mining, correlation network topology analysis and pattern recognition techniques (see Zhang et al. (2009) for a comprehensive review of the types of 'omics data that were integrated). More recently, multivariate projection-based methodologies have attracted a lot of attention as they have many advantages. First, they are computationally efficient to handle large datasets, where the number of biological features is much larger than the number of samples. Second, they perform dimension reduction by projecting the data into a smaller subspace while capturing the largest sources of variation from the data, resulting in a visual snapshot of the system under study. Third, they can be very flexible to address different types of biological problems. However, the projection of the data into a smaller subspace is often not sufficient to extract relevant information and answer the question, *"Which genes and proteins act in concert in the biological system under study?"*. Multivariate projection-based methods provide a framework for "feature selection", that is, the identification of relevant molecular features in the method.

In recent years, several variants based on partial least square (PLS) regression (these initials can also refer to projection to latent structure corresponding to the same method) and canonical correlation analysis (CCA) have been proposed to perform feature selection (Chung and Keles 2010; Lê Cao et al. 2009; Parkhomenko et al. 2009; Witten et al. 2009; Waaijenborg et al. 2008; Lê Cao et al. 2008). These methods focus on the integration of two 'omics datasets. The framework was extended to the integration of multiple datasets, first with Witten and Tibshirani (2009), who proposed to concatenate all datasets with an appropriate weight applied to each dataset, and then with Tenenhaus and Tenenhaus (2011), who extended the PLS framework with the regularized generalized canonical correlation analysis (rGCCA). More recently, Tenenhaus et al. (2014) proposed a subsequent sparse variant (sGCCA), which enables feature selection in each 'omics dataset. Singh et al. (2019) extended this framework to address the identification of key molecular biomarkers from multi-'omics assays in a supervised framework. An illustration of the efficiency of this approach is given in Lee et al. (2019).

In this chapter, we review some popular and recent multivariate methodologies for the exploration and integration of 'omics datasets. We first start with some generic mathematical background, then describe, in detail, multivariate approaches to explore one dataset (principal component analysis (PCA)), classify samples (projection to latent structure-discriminant analysis), integrate two datasets (projection to latent structure and related methods) and integrate several datasets (multi-block approaches). For each method introduced, we state the biological and statistical question, then illustrate the approach on a toy example based on simulated data. The remaining sections are dedicated to the interpretation of the results based on graphical outputs, then on a case study application studying liver toxicity in rats. The analyses are performed using the `mixOmics` R package (Lê Cao et al. 2009; González et al. 2009), which provides a framework for 'omics data exploration and integration. The R script used for toy examples and the real dataset is provided in the Appendix.

## 7.2. Background

### 7.2.1. *Mathematical notations*

We will use the following notations: $X$ denotes a data matrix of size $n \times p$ (e.g. transcriptomics), $Y$ denotes a data matrix of size $n \times q$ (e.g. other 'omics or clinical data), $n$ is the number of samples (e.g. mice, rats and plants) in rows, and $p$ and $q$ are the number of variables (transcripts and clinical measurements) in columns in X and Y, respectively. We will denote by $X'$ the transpose of $X$, which is of size $p \times n$. Unless specified, all methodologies center and scale each column in the data internally.

### 7.2.2. Terminology

We first describe the following terminologies used throughout this chapter:

– *Individuals, observations or samples*: these are the experimental units on which information are collected, for example, patients, cell lines, cells, biological samples and so on.

– *Variables and predictors*: these are read-outs measured on each sample, for example, gene (expression), protein or OTU (abundance), weight and so on.

– *Variance*: this measures the spread of one variable. In our methods, we estimate the variance of components rather that variable read-outs. A high variance indicates that the data points are spread out from the mean and from one another (scattered).

– *Covariance*: this measures the strength of the relationship between two variables, that is, whether they covary. A high covariance value indicates a strong relationship, for example, weight and height in individuals frequently vary roughly in the same way; roughly, the heaviest are the tallest. A covariance value has no lower or upper bound.

– *Correlation*: this is a standardized version of the covariance that is bounded by -1 and 1.

– *Linear combination*: here, variables are combined by multiplying each of them with a coefficient and adding the results. A linear combination of height and weight could be 2 × weight − 1.5 × height with the coefficients 2 and -1.5 assigned with weight and height, respectively.

– *Component*: this is an artificial variable built from a linear combination of the observed variables in a given dataset. Variable coefficients are optimally defined based on some statistical criterion. For example, in PCA, the coefficients in the (principal) component are defined so as to maximize the variance of the component.

– *Loadings*: variable coefficients used to define a component.

### 7.2.3. Multivariate projection-based approaches

Multivariate projection-based approaches, also called linear multivariate approaches, are powerful and particularly valuable approaches for high-throughput data, as they aim to reduce the dimension of the original data while highlighting the largest source of information from the data. In the remainder of this chapter, we will focus on multivariate approaches, which project the data into a small subspace spanned by latent variables (e.g. variables that are not directly observed) to reduce the dimensionality of the data. These multivariate methods are linear since the latent variables are defined according to a linear combination of the original molecular feature expression values. The terminology of the resulting vectors varies depending on the methodologies. For example, latent variables are called components in some of the approaches.

### 7.2.4. A criterion to maximize specific to each methodology

Linear multivariate approaches decompose a matrix, or a product of matrices, into a set of vectors called components (or latent variables) and loading vectors, where each component is associated with a loading vector. The matrix decomposition is usually efficiently performed using singular value decomposition (SVD (Alter et al. 2000)), which is the favored algorithm in most statistical software. For example, in PCA, we maximize the variance of the dataset $X$ by calculating the SVD of $X$. Canonical correlation analysis maximizes the correlation between two datasets $X$ and $Y$ with an eigen decomposition. PLS methods maximize the covariance between two datasets $X$ and $Y$ to extract related information between datasets by calculating the SVD of the matrix product $X'Y$. Generalized canonical correlation analysis maximizes the sum of covariances between pairs of $m$ datasets $X_1, X_2, \ldots, X_m$ (matching on the same samples $n$) to integrate those datasets by calculating the SVD of $X'_k X_l, k \neq l$, $k = 1, \ldots m$.

### 7.2.5. A linear combination of variables to reduce the dimension of the data

Since each sample can be characterized by $p$ (respectively, $q$) molecular features in each 'omics dataset $X$ (respectively, $Y$), we consider that each dataset "lives" in a space of dimension $p$ (respectively, $q$), where $p$ and $q$ can be very large depending on the type of 'omics that is considered. Each component is defined as a linear combination of the original molecular features from each 'omics dataset. In other words, each sample can be summarized by a numerical value (score), which is the sum of the expression value of each molecular feature multiplied by a particular weight. These weights (also called "coefficients") are directly obtained from the associated loading vector. The number of components must be specified by the user and can be set to a small value (usually less than 3 except for some complex problems), as we assume that the maximum information should be summarized by the first few components. The component represents a new (smaller) subspace where the original data are then projected and used for sample representations. The component values, or scores, provide insightful visualization to understand and visualize similarities between samples. A schematic view of loadings and components is provided for each method mentioned (see Figures 7.1, 7.3, 7.5 and 7.8).

### 7.2.6. Identifying a subset of relevant molecular features

When dealing with large-scale 'omics datasets, projecting the data in a smaller subspace spanned by a very small number of components is not enough to extract relevant information from the data. One of the reasons is that most of the molecules that are measured are not relevant to the biological study and are considered "noisy". During the last decade, a range of sparse multivariate approaches have been proposed to select the best subset of molecules relevant to the phenotype or biological outcome of interest. These sparse methods are based on the penalization of the loading vectors, so that most of the coefficients in the linear combination are shrunk to zero.

Therefore, the components are only based on a very small subset of relevant molecules. The idea is that the expression values of only relevant molecules that are combined in a linear manner should give insight into the data, while most of the "noisy" molecules are not taken into account when calculating the components. One of the most popular penalization approaches is based on the Lasso (least absolute shrinkage and selection operator) from Tibshirani (1996). This is the main feature selection approach that we will consider in this chapter, but note that other approaches also exist (such as Elastic Net, Fused Lasso, etc.). Feature selection is performed internally in the multivariate method, but requires the number of features to be selected on each component to be chosen. This can be arbitrarily chosen by the user (i.e. 10 genes on component 1, 25 genes on component 2, etc.) or by using other criteria based on a cross-validation strategy.

### 7.2.7. Summary

Linear multivariate approaches are powerful tools to project the data into a small subspace where similarities between samples can be visualized. The new subspace is defined by a small number of components and is a linear combination of the original variables. The weights in the linear combinations are determined by the associated loading vectors. These sets of vectors are obtained from a matrix decomposition of the original data matrices. Further insight in the data can be obtained via sparse Lasso penalized multivariate approaches, which enable the identification of relevant biological molecules from the data. In our framework, the user determines the number of variables to select on each component, rather than specifying a Lasso penalization parameter.

## 7.3. From the biological question to the statistical analysis

In this section, we start with a biological question that is then formulated into statistical terms. Mathematical and practical aspects of the methods are then discussed, and a summary is presented as a schematic view of the method. To conclude, a toy example based on simulated data illustrates the main aspects of the method.

### 7.3.1. *Exploration of one dataset: PCA*

#### 7.3.1.1. *Biological question*

I would like to identify the trends or patterns in my data or identify whether my samples "naturally" cluster according to the biological conditions. In addition, I would like to select the variables that contribute the most to the variability in the dataset.

#### 7.3.1.2. *Statistical point of view*

The question sets an unsupervised framework. Only numerical variables are taken into account, and no categorical information is provided to assign samples to known classes. A way to address this question consists of providing a visualization tool performing a reduction of the dimension of the dataset. PCA is dedicated to such a purpose. Selecting the most relevant variables can be achieved using a sparse version of PCA.

#### 7.3.1.3. *Methods*

##### 7.3.1.3.1. Mathematical aspects

The objective function to solve is

$$\arg\max_{||a^h||=1} \operatorname{var}(Xa^h) \quad h=1,\ldots,r \qquad [7.1]$$

where $X$ is the data matrix and $a^h$ is the $p$-dimensional loading vector associated with the principal component (PC) $h, h = 1, \ldots, r$ under the constraints that $a^h$ is of unit (norm) 1 and is orthogonal to the previous loading vectors $a^m, m < h$. The PC vector is defined as $t^h = Xa_h$.

Equation [7.1] can be solved by computing the eigenvectors $a_1, \ldots, a_r$ and associated eigenvalues $\delta_1, \ldots, \delta_r$ of the variance-covariance matrix $X'X$. The eigenvalues are ordered $\delta_1 \leq \delta_2 \leq \cdots \leq \delta_r$ and are proportional to the amount of explained variance on each dimension (component).

A more efficient way is to use the singular value decomposition of the data matrix $X$ ($X = U\Delta A'$), where the columns of $U$ are orthogonal and norm 1, and the columns of $A$ are the loading vectors. The matrix $\Delta = \mathrm{diag}(\delta_1, \ldots, \delta_r)$ contains the singular values (the square roots of the eigenvalues of $X'X$). The PCs are the columns of $T = U\Delta$. Computing the SVD of $X$ is more efficient than calculating the matrix product $X'X$, which is of size $p \times p$. Nowadays, the SVD is the standard way to calculate PCA.

A sparse version of PCA can be obtained using a Lasso penalty. It consists of adding the constraints $||a^h||_1 \leq \lambda_h$ to the expression to be maximized in equation [7.1], where $||a^h||_1 = \sum_{i=1}^{p} |a_i^h|$ is the $L^1$–norm of $a^h$ and $\lambda_h$ is a non-negative parameter that controls the amount of shrinkage in $a^h$. This results in a sparse loading vector and therefore in a selection of variables.

### 7.3.1.3.2. Practical aspects

*Choosing the number of dimensions*: as mentioned in section 7.2.4, PCA aims to maximize the variance in the dataset via the eigen decomposition of the matrix $X'X$ or equivalently via the SVD of the matrix $X$. The result from the matrix decomposition is a set of components (called PC) for which their variance is maximized. In PCA, the variance is equivalent to the information that can be extracted from the data. These PCs are ranked with decreasing variance. In PCA, we say that a PC "explains a certain amount of variance in the data", which corresponds to the variance of the PC. This is usually visualized via a variance barplot or scree plot. The barplot is the most visual ad hoc way to choose the final number of PCs to retain in the analysis. For example, we should choose the number of PCs indicated by the "elbow" in the barplot (which means that the following PCs explain very little variation in the data) or if the proportion of variance explained by the first PCs is considered enough to summarize the information in the data. Depending on the type of 'omics data, 60% of the total variation is often considered sufficient. PCA is useful to visualize the similarities between samples once the data are projected onto the PCs. The sample plot can often highlight unknown sources of variations, for example, due to hospital locations, when we least expect it. It can also assess the "potential" in the data, providing a tentative answer to the question: how easy or difficult can it be to separate the different treatment groups? Finally, the sample plot can also bring out possible sample outliers.

*Limitations*: since PCA aims to maximize the variance, the presence of any sample outlier can inflate the amount of variance explained. It may be useful to go back to, for instance, the clinical information of the problematic sample and assess whether the sample in question should be flagged as an outlier and removed. Obviously, removing samples considered as outliers should be done with caution. Furthermore, PCA assumes that all samples are independent, that is, PCA cannot take into account a cross-over design where the patient acts as their own control (e.g. before/after treatment). Recent developments have been proposed with multilevel

PCA (de Noord and Theobald 2005; Velzen et al. 2008) to account for the individual variability in the data. PCA applied wrongly to a study where patients go through several treatments after a wash-out period can highlight a strong individual effect (i.e. the samples from the same patient are clustered) rather than a treatment effect. Finally, PCA is often not very informative when the number of molecular features is very large in the dataset, as irrelevant features bring noise and blur the important signal. One solution is to prefilter the data, for example, removing undetected probes below a specified threshold in a microarray experiment or removing molecular features with a very small variance across all samples or, in the same vein, keeping the molecular features with the highest variance. Another solution consists of applying a sparse approach that selects the relevant biological features in the PCA, according to the maximization of the data variance.

*Related methods*: the biological question may not be related to the highest variance. In that case, independent component analysis (ICA (Comon 1992; Hyvärinen and Oja 2000)) may be more appropriate. While PCA requires the PCs to be orthogonal (i.e. uncorrelated), ICA requires the components to be statistically independent, which means that the values of one component provide no information about the values of other components. The assumption of ICA is that the variables that we measure depend on some biological or environmental factors that are assumed to be statistically independent. The "independent components" we are seeking in ICA correspond to those factors. Due to its independence condition, ICA may be more suitable to some metabolomics studies and outperformed PCA (Scholz et al. 2004; Wienkoop et al. 2008). ICA can be performed in R using the packages `ica` (Helwig 2018) and `fastICA` (Marchini et al. 2019). More recently, Yao et al. (2012) proposed a variant of ICA called independent principal component analysis (IPCA) that uses ICA as a denoising process of the PCA loading vectors; a sparse version for variable selection was also proposed with sparse IPCA, and both are implemented in the mixOmics package (Rohart et al. 2017a).

### 7.3.1.3.3. Summary

Figure 7.1 provides a schematic view of the decomposition of the dataset into components using loading vectors for the PCA (A) and the sparse PCA (B).

### 7.3.1.4. *Toy example*

To illustrate the principle of PCA, we simulated three datasets, each composed of 50 observations and three quantitative variables. In the first dataset, the three variables are simulated with nearly zero pairwise correlations (Figure 7.2(A)). In the second dataset, two variables are simulated with a strong correlation (Figure 7.2(B)). In the third dataset, the three variables have strong correlations (Figure 7.2(C)).

The percentage of variance explained by the PCs on each of the three datasets are presented in Table 7.1. We can conclude the following:

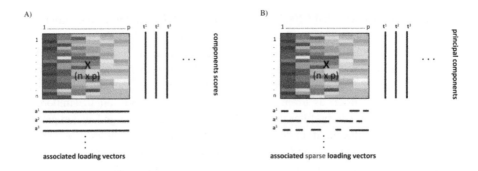

**Figure 7.1.** *Illustration of the matrix decomposition of the matrix $X$ into a set of vectors: three components $t^1, t^2, t^3$ and associated loading vectors $a_1, a_2, a_3$ that are obtained via the singular value decomposition of $X$. The component $t^1$ is associated with the loading vector $a_1$ and so on for (A) classical principal component analysis and (B) sparse principal component analysis, where elements in the loading vectors are set to zero*

| Dataset 1 | PC1 | PC2 | PC3 |
|---|---|---|---|
| St. Dev. | 0.32 | 0.30 | 0.27 |
| Prop. of Var. | 0.38 | 0.34 | 0.28 |
| Cum. Prop. | 0.38 | 0.72 | 1.00 |

| Dataset 2 | PC1 | PC2 | PC3 |
|---|---|---|---|
| St. Dev. | 1.51 | 0.93 | 0.35 |
| Prop. of Var. | 0.70 | 0.27 | 0.04 |
| Cum. Prop. | 0.70 | 0.96 | 1.00 |

| Dataset 3 | PC1 | PC2 | PC3 |
|---|---|---|---|
| St. Dev. | 1.91 | 0.43 | 0.31 |
| Prop. of Var. | 0.93 | 0.05 | 0.02 |
| Cum. Prop. | 0.93 | 0.98 | 1.00 |

**Table 7.1.** *Standard deviation (St. Dev.), proportion of variance explained (Prop. of Var. in %) and cumulative proportion of variance explained (Cum. Prop. in %) by the three principal components in each simulated dataset*

– **Dataset 1:** the variance explained by each of the three PCs is nearly equivalent (around 30%) as no specific shape can be found into the locations of the observations in the three-dimensional space.

– **Dataset 2:** the variance explained by the third PC is negligible (lower than 4%) as the information contained in the dataset can be summarized in two dimensions.

– **Dataset 3:** only the first PC contains relevant information (93%) as the observations follow a straight line, even in a three-dimensional space.

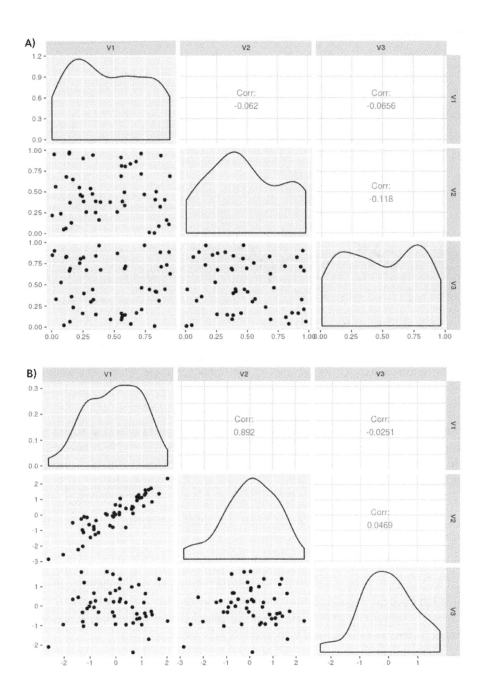

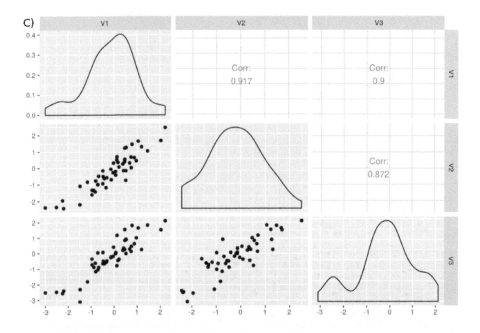

**Figure 7.2.** *View of the three simulated datasets with 50 observations and three variables, from top to bottom: (A) no correlation between variables; (B) V1 and V2 are highly correlated; (C) V1, V2 and V3 are highly correlated. Plots were produced with the* GGally *package (Schloerke et al. 2018)*

In the second and third datasets, the results of PCA reveal that we can summarize the data in a smaller sub-space (of dimension 2 and dimension 1, respectively). To achieve this, PCA takes benefit from the redundancy provided by highly correlated variables.

### 7.3.2. Classify samples: projection to latent structure discriminant analysis

#### 7.3.2.1. Biological question

I wish to analyze one dataset. I am interested in classifying my samples into known classes and would like to know whether my numerical data can rightly classify the samples, as well as predict the class of new samples. In addition, I would like to select a small number of variables that help to classify the samples.

#### 7.3.2.2. Statistical point of view

The question clearly sets a supervised framework, as a categorical outcome is considered to classify samples into known classes. To answer this question, we need to handle a matrix of numerical variables and a categorical outcome. Fisher's discriminant analysis (also called linear discriminant analysis) is one of the most commonly used multivariate discriminant analysis approaches, but is limited in a large dataset setting. The high number of variables creates computational issues to estimate the inverse of a large variance–covariance matrix. One solution we consider here is partial least squares discriminant analysis (PLS-DA (Barker and Rayens 2003; Chung and Keles 2010)), which circumvents the matrix inversion issue with the use of latent variables. PLS-DA is a dimension reduction technique that aims to discriminate known sample groups. Once the model is fitted, prediction performances can be assessed. A sparse variant has also been proposed to select the relevant discriminative features from the dataset (Lê Cao et al. 2011).

### 7.3.2.3. Method

#### 7.3.2.3.1. Mathematical aspects

The objective function to solve is

$$\arg\max_{||a^h||=1, ||b^h||=1} \mathrm{cov}(Xa^h, Y^*b^h) \quad h = 1, \ldots, r \qquad [7.2]$$

where $Y^*$ is the categorical response coded as a dummy matrix, which includes a column for every group recorded in the outcome vector $y$, whereas $t = Xa$ and $u = Y^*b$ are the PLS-DA components.

PLS-DA maximizes the covariance between the data matrix $X$ and the dummy matrix $Y^*$. The components that are obtained from the SVD of $X'Y^*$ are called latent variables in the PLS terminology. These latent variables maximize the discrimination between classes.

#### 7.3.2.3.2. Practical aspects

*Tuning the method*: the method requires us to choose the number of components (usually a small number) and the number of features to select on each component for the sparse version. The latter can be determined using cross-validation to assess the optimal number of features that will give the best performance (i.e. the lowest classification error rate). We usually advise choosing $K - 1$ components, where $K$ is the number of classes. However, our experience has shown that adding more components can be beneficial to increase the performance of the model.

*A predictive statistical model*: as a multivariate approach, PLS-DA is particularly adapted to large datasets and is not limited to problems with only two classes. The latent variable scores are defined so that each linear combination of the original variables (using the coefficients in the loading vectors) can distinguish the class of each sample in what we call the training set. To predict the class of a new sample (from an external test set), we calculate the linear combination of the new sample using the expression values of the $p$ molecular features multiplied by the coefficients indicated in the loading vector. The score indicates the predicted class of each sample. An example of a full case study where a training and a test set were used for prediction and validation can be found in Günther et al. (2014).

*Feature selection bias*: feature selection bias is a common pitfall to avoid. The number of features to select on each component is the main parameter to choose in the method. In the early days of microarray data, studies often reported a very optimistic classification error rate close to 0% for a given number of selected genes. Ambroise and McLachlan (2002) demonstrated that these studies suffered from the "feature selection bias" issue, or overfitting, which occurs when the performance of the model is tested on the same samples on which the classification rule has been defined. In other words, the features have been selected based on their good performance on both the training *and* the test set. The problem often arises when there is no external test set to assess the performance of the model. In that case, we have to resort to cross-validation, which partitions the samples into training and test sets multiple times, with the prediction error averaged across all partitions. A classification error rate that is too optimistic is often obtained when, despite cross-validation, the features have been selected on all samples beforehand and then re-tested on the same samples in a "test" set. The problem of overfitting is now well-recognized (and feared for) in the 'omics data analysis field. We measure overfitting by evaluating the statistical model on an external test set, which preferably includes a number of samples superior to those in the training set. However, it is often impractical or too expensive to generate such an external test set. Further, using an independent study as an external test set can lead to the problem of a batch effect (or systematic bias) that would need to be corrected first. In summary, when evaluating the performance of the model given a selection of features, we must ensure that the selection does not overfit the data by using careful cross-validation or external test sets when possible.

*Limitations*: similar to PCA, PLS-DA considers that samples are all independent. In the case of repeated measures or cross-over designs, a multilevel variant has been proposed by Westerhuis et al. (2010) and Liquet et al. (2012) to accommodate for this particular type of experimental design. Another limitation is the small number of samples in one (or more) treatment groups, which can result in bias during performance evaluation. First, cross-validation involves partitioning the samples, leading to classes that may not be represented in the training or test set. Therefore, it is important to ensure that the original proportion of samples per group is maintained, provided that there is a sufficient number of samples per group ($\geq 3$). Our functions in mixOmics include stratified sampling. Second, when the groups have an unbalanced number of samples, the calculation of the overall classification error rate can be biased toward the class that has the highest number of samples. In that case, it is preferable to weight the classification error rate with respect to the number of samples per class (also called "balanced classification error rate") or report the classification error rate per class.

### 7.3.2.3.3. Summary

Multivariate discriminant analysis seeks optimal components that best discriminate the classes of the categorical outcome of interest. The approach requires as an input the data matrix (e.g. expression dataset $X$) and the information about the group of each individual denoted $y$. The vector $y$ is converted into a dummy matrix $Y^*$ composed of 0 and 1 and with as many columns as the number of categories (Figure 7.3).

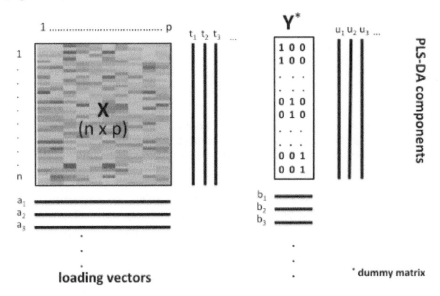

**Figure 7.3.** *Illustration of the matrix decomposition of the matrix $X$ into a set of vectors: three components $t^1, t^2, t^3$ and associated loading vectors $a_1, a_2, a_3$ that are obtained by maximizing the covariance with a set of components $u$ coming from the decomposition of the dummy matrix $Y^*$ standing for the factor $y$*

The sparse PLS-DA could be represented in the same way as sparse PCA in Figure 7.1 with sparse loading vectors.

### 7.3.2.4. Toy example

The toy example used to illustrate a discriminant analysis is composed of three quantitative variables and one categorical variable for 50 observations. Figure 7.4

illustrates variables V1 and V2 with a higher variance (respectively, 84 and 32) than V3 (2), but no difference can be seen between the two groups. Conversely, V3 has a smaller variance, but the values are different depending on the two groups.

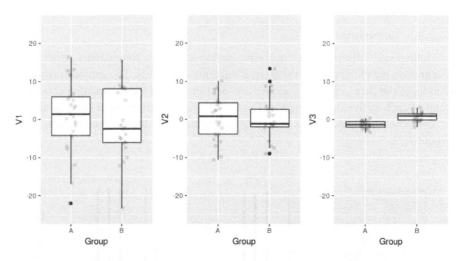

**Figure 7.4.** *Boxplots by group of the three variables (V1, V2 and V3 from left to right) in the simulated dataset. The variances of the variables V1, V2 and V3 are 84, 32 and 2, respectively*

PCA on this dataset restricted to V1-V3, without taking into account the group information, results in a first PC very similar to V1, as its variance is much greater than the others. PC1 is defined as: $PC1 = -0.99 \times V1 - 0.008 \times V2 - 0.003 \times V3$ and thus maximizes the variance of the observations projected on PC1.

On the contrary, a discriminant analysis to classify samples from groups A and B gives more importance to V3. Hence, the first PLS-DA component is defined as: $PLS\text{-}DA1 = -0.088 \times V1 + 0.034 \times V2 - 0.99 \times V3$. Note that the coefficient associated with $V3$ is more than 10 times greater than the other two.

This toy example emphasizes the importance of clearly formulating the nature of the unsupervised or supervised problem.

### 7.3.3. Integration of two datasets: projection to latent structure and related methods

#### 7.3.3.1. *Biological question*

I would like to know whether I can extract common information from two datasets measured on the same samples or highlight their correlation. In addition, I would like to select the variables from both datasets that covary (i.e. "change together") across the different conditions.

#### 7.3.3.2. Statistical point of view

The questions clearly set an integration framework based on either covariance or correlation.

Canonical correlation analysis (Hotelling 1936) maximizes the correlation between two matching datasets (e.g. transcriptomics and metabolomics data measured on the same samples). In the same vein as PCA, CCA seeks linear combinations of the variables (called canonical variates) to reduce the dimension of the datasets, while maximizing the correlation between the two variates (the canonical correlations).

Partial least square regression (Wold et al. 2001), also known as projection to latent structures, generalizes PLS-DA to the case where two datasets contain continuous measurements on the same samples. Here, the information from both datasets is integrated in an unsupervised manner (no information about the patients class or phenotype is included in the PLS model), and PLS thus differs from PLS-DA. PLS methods are very powerful and flexible tools to address a variety of biological questions (see Boulesteix and Strimmer (2007); Esposito Vinzi et al. (2010) for a review).

#### 7.3.3.3. Methods
#### 7.3.3.3.1. Mathematical aspects

Like PCA and PLS-DA, CCA and PLS are multivariate linear methods optimizing a specific criterion.

$$\arg\max_{||a^h||=1, ||b^h||=1} \text{cor}(Xa^h, Yb^h) \quad \text{for CCA} \qquad [7.3]$$

$$\arg\max_{||a^h||=1, ||b^h||=1} \text{cov}(Xa^h, Yb^h) \quad \text{for PLS} \qquad [7.4]$$

There are three major differences between PLS and CCA. First, the PLS algorithm uses the latent variables as a mean to avoid the inversion of large data matrices, and is therefore not limited by a large number of features, while CCA requires the calculation of the inverse of the variance–covariance matrix that can be ill-defined in the large dimensional case. Second, the matrix decomposition in PLS is performed in an iterative manner (for each step, the information extracted and summarized from the previous PLS component is removed from the current data matrix), which allows for the modeling of the data structure. Third, the PLS framework enables the inclusion of penalization parameters (Lasso) to perform variable selection and thus identify molecular signatures from both datasets.

#### 7.3.3.3.2. Practical aspects

*Limitations of CCA*: a major limitation of CCA is the high dimensionality of the data. When the number of variables increases, CCA results in several canonical correlations close to 1, indicating that the canonical subspace cannot uncover any meaningful association. Another limitation of CCA is that it requires the computation of the inverse of ill-conditioned matrices, that is, matrices with highly correlated variables within each dataset. This leads to unreliable matrix inverse estimates. One solution is to include a regularization step in the CCA calculation by introducing constants, also called Ridge penalties, on the diagonal of the variance–covariance matrices of $X$ and $Y$ to make them invertible. The choice of the penalty parameters $(\lambda_1, \lambda_2)$ in regularized CCA (rCCA) can be guided with cross-validation to obtain stable canonical variates, as proposed by González et al. (2008), and it was shown to give biologically meaningful results (Combes et al. 2008). A set of canonical variates for each dataset is obtained. The common way to represent the samples is to project the samples in the subspace spanned by the mean of the two canonical variates associated with the $X$ and $Y$ datasets. The mixOmics package also implements a direct estimation of the parameters based on the shrinkage approach from Schäfer and Strimmer (2005).

*Sparse PLS and the selection of co-expressed features*: several sparse PLS (sPLS) variants have been proposed (Waaijenborg et al. 2008; Lê Cao et al. 2008; Parkhomenko et al. 2009; Witten et al. 2009; Chun and Keles 2010). The differences between these approaches rely on algorithmic features or the penalization term. The sPLS that we consider includes Lasso penalization on the loading vectors associated with both datasets (Lê Cao et al. 2008), which enables the selection of 'omics features in each dataset that are co-expressed or correlated (positively or negatively) on each PLS dimension. In this unsupervised setting, choosing the optimal number of features to select remains an open question. We favor the use of a stability criterion (implemented in the mixOmics package), which records the features that are repeatedly selected when we partition the samples into several folds. The stability measure represents a frequency of selection for each feature when the original data are perturbed. We refer the reader to Günther et al. (2014) for a detailed analysis on a study with transcriptomics and metabolomics data.

#### 7.3.3.3.3. Summary

Multivariate integrative analyses seek optimal coefficients that optimally associate each of the two datasets (Figure 7.5) by maximizing either covariance (PLS) or correlation (CCA) between components.

The sPLS could be represented in the same way as sparse PCA in Figure 7.1 with sparse loading vectors for one or both datasets.

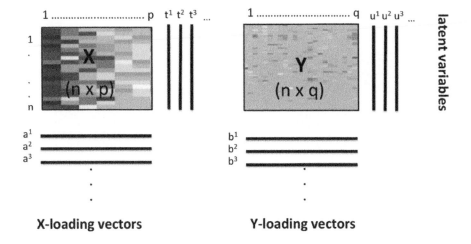

**Figure 7.5.** *Illustration of the matrix decomposition of the matrices $X$ and $Y$ into two sets of vectors: three components $t^1, t^2, t^3$ and associated loading vectors $a_1, a_2, a_3$ associated with $X$. The loading vectors are obtained by maximizing the covariance or the correlation with a set of components $u^1, u^2, u^3$ with loading vectors $b_1, b_2, b_3$ associated with $Y$*

### 7.3.3.4. Toy example

We simulated a dataset of 50 observations of five variables in X and three variables in Y. Variables $X_1$ and $Y_1$ are positively correlated (0.8); variable $Y_3$ is negatively correlated with $X_2$ (-0.7) and positively correlated with $X_3$ (0.6). Other correlations are lower than 0.4. Figure 7.6(A) provides a global view of the correlation structure within and between X and Y.

The variable representation of a PLS analysis is presented in Figure 7.6B. The interpretation of a correlation circle plot is briefly discussed in section 7.4 and is detailed in González et al. (2012). This plot clearly exhibits the expected correlation between our variables of interest. The opposition between $X_2$ and $(X_3, Y_3)$ defines the first canonical dimension (horizontal), which is consistent with the relationships between these three variables. The second canonical dimension (vertical) highlights the positive correlation between $X_1$ and $Y_1$. Other variables are located near the origin of the plot indicating weak correlation. These variables would certainly be removed if sPLS was used.

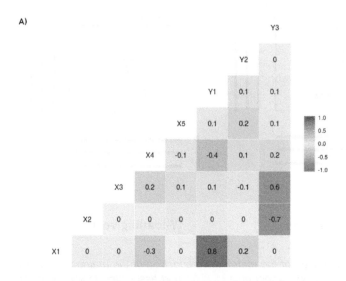

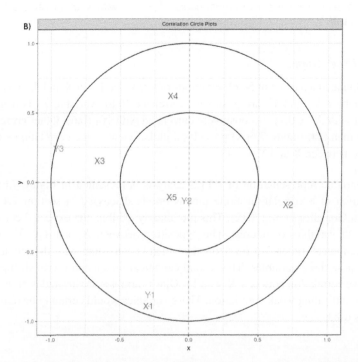

**Figure 7.6.** *(A) Correlation structure between $X$ and $Y$ datasets. (B) Variable representation on the first two PLS components in a correlation circle plot*

### 7.3.4. Integration of several datasets: multi-block approaches

#### 7.3.4.1. Biological question

I would like to understand the relationship between more than two datasets and identify a highly correlated multi-'omics signature discriminating groups of samples.

#### 7.3.4.2. Statistical point of view

The question expands our previous PLS/CCA biological issue addressed in section 7.3.3 by considering more than two numerical datasets and, potentially, a categorical outcome. This case is sometimes referred as N-integration, as every dataset is acquired on the same $n$ or $N$ samples. To answer this question, a generalized version of the CCA/PLS methods (GCCA) need to be used. In addition, the selection of the most relevant variables in each dataset can be achieved using a sparse penalized version of the method. As the complexity of the biological question (and the data) increases, so does the computational method. In this case, we will need to include the strength of the assumed correlation between pairs of datasets in the statistical model.

#### 7.3.4.3. Method

##### 7.3.4.3.1. Mathematical aspects

We consider a general framework based on the concepts we introduced in section 7.3.3 for PLS and CCA, where the criterion to maximize is the covariance or the correlation between pairs of datasets. To formalize this framework, we have to adopt new notations. We denote as $Q$ the 'omics datasets $X^{(1)}(n \times p_1)$, $X^{(2)}(n \times p_2),...,X^{(Q)}(n \times p_Q)$ measuring the expression levels of $P_q$ 'omics variables on the same $n$ biological samples. GCCA solves for each component $h = 1,\ldots,H$:

$$\max_{a_h^{(1)},...,a_h^{(Q)}} \sum_{q,j=1, q \neq j}^{Q} c_{kq} \text{cov}(X_h^{(q)} a_h^{(q)}, X_h^{(j)} a_h^{(j)}) \qquad [7.5]$$

$$\text{subject to} \quad ||a_h^{(q)}||_2 = 1 \quad \text{and} \quad ||a_h^{(q)}||_1 \leq \lambda^{(q)}$$

where $\lambda^{(q)}$ is the penalization parameter, $a_h^{(q)}$ is the loading vector on component $h$ associated with the residual (deflated) matrix $X_h^{(q)}$ of the dataset $X^{(q)}$ and $C = \{c_{q,j}\}_{q,j}$ is the design matrix that specifies whether datasets should assumed to be correlated (see the discussion about the design matrix in the practical aspects). Details about GCCA can be found in (Tenenhaus and Tenenhaus 2011; Tenenhaus et al. 2014;

Singh et al. 2019). Other frameworks exist for multi-block integration: for instance, 16 methods published between 1961 and 2006 are listed in Tenenhaus and Hanafi (2010). They are not necessarily adequate for large 'omics data and would need to be further adapted for such a context.

### 7.3.4.3.2. Practical aspects

The practical aspects regarding variable selection and the choice of the dimension that have been previously mentioned for the other methods (sections 7.3.1, 7.3.2 and 7.3.3) are still valid for the multi-block approaches. They are discussed in the multi-blocks framework in the supplementary information by Rohart et al. (2017a). To avoid repetitions, we focus here on the design matrix that is specific to these multi-blocks analyses with GCCA.

*Design matrix*: the elements in the design matrix $C$ can be set to values ranging from 0: null design, datasets are not connected, to 1: full design, datasets are fully connected. Figure 7.7 illustrates three different scenarios, including an intermediate (full weighted design) between these two extreme cases.

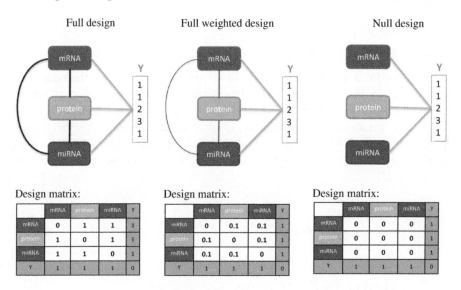

**Figure 7.7.** *Example of different design matrices in DIABLO for the multi-'omics breast cancer study presented in Rohart et al. (2017b). Links or cells in gray are added by default in the supervised context and do not need to be specified by the user*

The design matrix can be defined based on biological knowledge (e.g. transcriptomics can be assumed to be more correlated to proteomics than to clinical variables) or a data-driven approach when two-'omics integration analyses have already been performed and interpreted.

### 7.3.4.3.3. Summary

When the aim is to integrate more than two datasets, the criterion to optimize must take into account every dataset in a paired combination, as illustrated in Figure 7.8. A design matrix needs to be specified to model or favor particular relationships. The multi-block methods seek components in each dataset that are maximally covariant. Each component is defined as a linear combination of variables from their own dataset.

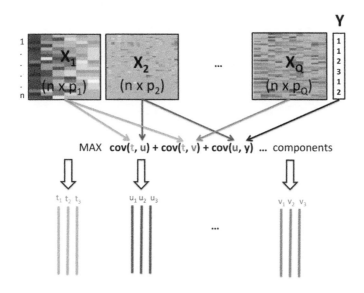

**Figure 7.8.** *Illustration of the matrix decomposition of several matrices $X_1, X_2, \ldots$ into sets of vectors: components $(t^1, t^2, t^3)$, $(u^1, u^2, u^3)\ldots$ with associated loading scores not represented here*

### 7.3.4.4. Toy example

The toy example related to multi-block analysis combines the simulated studies presented in PLS-DA (section 7.3.2) and PLS (section 7.3.3). It includes three numerical datasets $X$ (five variables), $Y$ (three variables) and $Z$ (eight variables) for 50 observations. The correlation matrix between each pair of variables is represented in Figure 7.9(A). Highly correlated pairs are $(X1, Y1)$, $(X3, Y3)$, $(X3, Z8)$, $(Y3, Z8)$ for positive correlations, and $(X2, Y3)$, $(X2, Z3)$ for negative ones. Furthermore, only $Z7$ discriminates the two sample groups A and B (Figure 7.9(B)).

218 Biological Data Integration

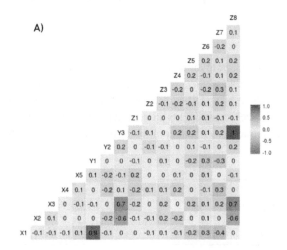

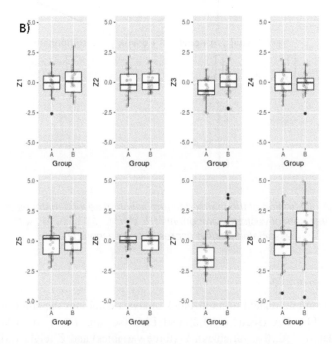

**Figure 7.9.** *(A) Correlations between every pair of variables from $X$, $Y$ and $Z$. (B) Boxplots with respect to sample group in $Z$. $Z7$ is the only discriminant variable between the sample groups A and B*

From the correlation circle plot (Figure 7.10(A)), the correlation structure between variables is adequately identified, as shown by the proximity between $X1$ and $Y1$ (bottom-left corner); between $X3$, $Y3$ and $Z8$ (bottom-right corner); and the opposite (negative) correlation of $X2$ with $X3$, $Y3$ and $Z8$.

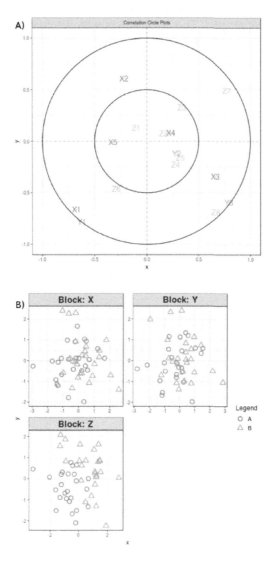

**Figure 7.10.** *Results of a multi-block PLS-DA. (A) Representation of the variables. (B) Representation of the individuals on the three subspaces spanned by the variables in each dataset*

From the sample plots (Figure 7.10(B)) obtained from each set of components associated with each dataset, a discrimination between the two groups of samples can be observed in the $Z$ subspace. This is because the $Z$ dataset includes the only discriminant variable $Z7$. It can also be noted that $Z7$ is located in the upper right corner on the correlation circle plot, as well as the observations from group B (triangles) on the sample plot. This is consistent with the greater values of samples from group B for $Z7$ (see Figure 7.9(B)).

## 7.4. Graphical outputs

The interpretation of the results of the methods presented in section 7.3 are mainly based on graphical outputs. A detailed explanation of several graphical outputs is provided in González et al. (2012). We provide here a basic description of the different types of graphics to visualize samples or variables. In addition, we mention the related functions implemented in the `mixOmics` to generate such graphical outputs.

### 7.4.1. Individual plots

Sample plots display the components, and therefore visualize similarities between samples in a reduced dimensional space spanned by the first few latent components of the model.

#### 7.4.1.1. Sample score representation

Scores representation can be displayed using the `plotIndiv` function. Examples can be seen in Figures 7.10(B), 7.11 and 7.13. For the integrative methods, samples from separate dataset are plotted by default on separate figures, allowing us to assess the agreement between the datasets at the sample level. For a two-block integration, one plot can be displayed that considers the subspace spanned by the mean of the components associated with the $X$ and $Y$ datasets. To ease the interpretation of sample groups, when they are known, confidence ellipses for each sample group can be displayed. Furthermore, with discriminant methods, users can overlay prediction results on sample plots via the background input parameter to visualize each class prediction area. The method defines surfaces around samples that belong to the same predicted class. These surfaces are then used to shade the background of the sample plot. See the `background.predict` function and the optional argument background in `plotIndiv`.

#### 7.4.1.2. Arrow representation

This plot can be obtained with the `plotArrow` function. It overlays the components scores from multiple datasets and draws arrows between scores associated with the same sample. For most supervised methods and two-'omics integration methods, the start of the arrow represents the component in the $X$ dataset and the tip of the arrow is the component associated with the $Y$ dataset, or the outcome in the case of supervised analysis. For N-integration, the start of the arrow indicates the centroid between all 'omics datasets for a given sample and the tip of the arrow the location of the same sample in each dataset. In two-'omics and N-integration methods, short arrows indicate a strong agreement between the matching datasets, and long arrows indicate a disagreement between the matching datasets. This representation is not illustrated in this chapter.

### 7.4.2. Variable plots

Variable plots aim to display the relationships between selected variables across datasets by using the latent components as a surrogate variable to estimate the correlations (correlation circle plots) or associations between variables (clustered image maps and relevance networks). The loading vectors plot displays the importance of each selected variable, and its contribution with respect to a sample group. Some plots display specified components (plotVar, plotLoadings), while others can also aggregate the similarities between variables across all components (cim, network).

#### 7.4.2.1. Correlation circle plots

Correlation circle plots display the correlation between variables (biological features) and latent components. Each variable coordinate is defined as the Pearson correlation between the original data and a latent component. Correlation circle plots are particularly useful to visualize the contribution of each variable to define each component (variable close to the large circle of radius 1), as well as the correlation structure between variables (clusters of variables). The cosine angle between the segments joining any two points to the origin represents the correlation (negative, positive or null) between two variables. Examples of correlation circle plots can be seen in Figures 7.6(B), 7.10(A) and 7.12.

#### 7.4.2.2. Loading plots

These plots displayed with the plotLoadings function represent the loading weights of each variable (selected) on each dimension of the multivariate model. Most important variables (according to the absolute value of their coefficients) are ordered from bottom to top. For supervised analyses, colors indicate the class for which the mean (or median) expression value is the highest or the lowest for each feature (contrib = ''max'' or ''min''). This graphical output enables us to better characterize the molecular signature, especially when interpreted in conjunction with the sample plot. Examples of loading plots can be seen in Figures 7.14 and 7.17(B).

### 7.4.2.3. Relevance networks

Relevance networks represent the correlation structure between variables of different types. They can be produced using the `network` function. The function avoids the intensive computation of Pearson correlation matrices on large datasets by instead calculating a pair-wise similarity matrix directly obtained from the latent components of the integrative approaches (CCA, PLS and multi-block methods). The similarity value between a pair of variables is obtained by calculating the sum of the correlations between the original variables and each of the latent components of the model. The values in the similarity matrix can be seen as a robust approximation of the Pearson correlation (see González et al. (2012) for a mathematical demonstration and exact formula). The advantage of relevance networks is their ability to simultaneously represent positive and negative correlations, which are missed by methods based on Euclidean distances or mutual information. Those networks are bipartite and thus only a link between two variables of different types can be represented. This representation is not used in this chapter.

### 7.4.2.4. Clustered image maps

The plot visualizes the distances between two types of variables (two-'omics integration) using the `cim` function. Clustered image maps (CIM) are based on a hierarchical clustering simultaneously operating on the rows and columns of the selected variables in the original data for the latter and on the similarity matrix defined in the network visualization for the former. By default, we use the Euclidean distance and complete linkage method, but other distances and methods are proposed. Examples of clustered image maps can be seen in Figure 7.20.

## 7.5. Overall summary

We summarize the different steps required to conduct the multivariate analyses introduced in this chapter. We also refer to the functions of the `mixOmics` package for each step:

1) formulate the biological question;

2) choose the appropriate method, using the functions: `pca`, `spca`, `plsda`, `splsda`, `pls`, `spls`, `block.plsda`, `block.splsda`;

3) tune or choose the parameters of the function, when applicable;

4) represent the individuals with sample plots: `plotIndiv`, `plotArrow`;

5) represent the variables with variable plots: `plotVar`, `plotLoadings`, `cim`, `network`;

6) interpret the results, answer the question and, potentially, return to step 1 with new questions that may have arisen from the analysis.

Section 7.6 illustrates the analysis of a real dataset in detail.

## 7.6. Liver toxicity study

### 7.6.1. *The datasets*

In the liver toxicity study, 64 male rats of the inbred strain Fisher 344 were exposed to non-toxic (50 or 150 mg/kg), moderately or severely toxic (1,500 or 2,000 mg/kg) doses of acetaminophen (paracetamol) in a controlled experiment. For simplicity reasons, we recoded this factor into two levels: low and high. Necropsy were performed at 6, 18, 24 and 48 h after exposure and the mRNA from the liver was extracted. The study is composed of two datasets and provides a good illustration of the usefulness of the multivariate projection-based methodologies presented in this chapter. The clinical dataset includes 10 clinical chemistry measurements of variables containing markers for liver injury. The serum enzyme levels were numerically measured for each subject (rat). The transcriptomic dataset contains the expression of 3,116 gene expression levels. The data were normalized and preprocessed by Bushel et al. (2007) and are available through the mixOmics R package Rohart et al. (2017a).

### 7.6.2. *Biological questions and statistical methods*

They are several types of questions related to this study. For each question, we identify the statistical method presented in this chapter that will be used to provide some elements of answers.

1) Based on the transcriptomics, and the clinical data, do we naturally observe clusters of samples that correspond to the different dose or exposure treatments?

⇒ PCA on transcriptomics data and on clinical data separately.

2) Based on transcriptomics data, can we identify a molecular signature that characterizes the different treatment doses? Do we observe better discrimination with the clinical data?

⇒ Sparse PLS-DA on the transcriptomics data using the factor "treatment dose". We can compare the results with PLS-DA run on the clinical data with the same categorical outcome.

3) Can we unravel relationships between transcriptomics data and clinical data? What are the genes that characterize these relationships?

⇒ PLS will provide clues regarding the potential relationships between transcripts and clinical variables. The selection of the most relevant transcripts involved in these relationships can be achieved with the sparse version of PLS.

4) Does the integration of the clinical and transcriptomics datasets bring better insight into the biological similarities between samples within the same treatment?

⇒ A multi-block approach is required to take into account the two matrices composed of numerical variables, as well as a categorical outcome. The selection of transcripts can be done using a sparse version.

### 7.6.3. *Single dataset analysis*

Prior to the integration of several 'omics datasets, it is crucial to have a first understanding of each dataset analyzed separately. Two types of analysis can be performed. First, unsupervised approaches do not take into account the class or group label of each sample and aim to summarize the data according to the criterion that the method has been designed to maximize. Second, supervised approaches seek the best linear combination of molecular features, so that the different groups of individuals are well separated.

#### 7.6.3.1. *Unsupervised analysis*

We applied PCA on the transcriptomics and clinical data separately. Table 7.2 displays the proportion of explained variance per PC for the transcriptomics and the clinical data. For the transcriptomic data, the first two PCs explained 43% of the total variation (63% for the clinical data), while adding a third PC only explained another 8% of the total variation (12.5% for the clinical data).

We only consider the first two PCs, which summarize most variation in the data. Figure 7.11 represents the samples projected on the first two PCs for each dataset. For both cases, the samples seemed to cluster according to the low and high doses primarily. The difference between the time of exposure is somewhat unclear. The proportion of explained variance that is higher in the clinical data than in the transcriptomics data suggests that more information (variance) could be extracted from the clinical dataset.

The representation of the variables is of little interest for transcriptomic data (3,116 data points need to be represented!). Instead, we present the results of a sparse PCA selecting 10 genes on each of the first two components.

The two plots in Figure 7.12 should be jointly interpreted with the sample representation (Figure 7.11). For instance, one specific gene (on the top) seems to be characteristic of samples with high dose and long exposure time (24 h). The names of the selected genes can be retrieved using the `selectVar` function.

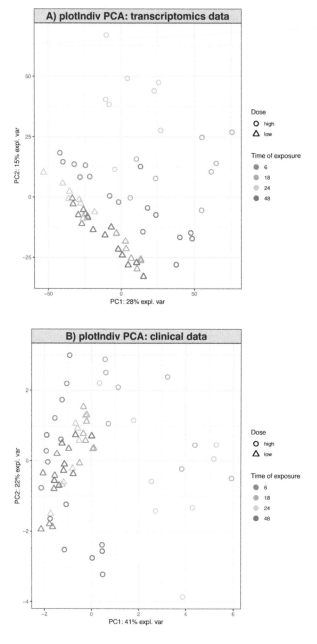

**Figure 7.11.** PCA on the transcriptomics and clinical data. Sample representation on each dataset: (A) transcriptomics, (B) clinical – analyzed separately when the data are projected on the first two principal components

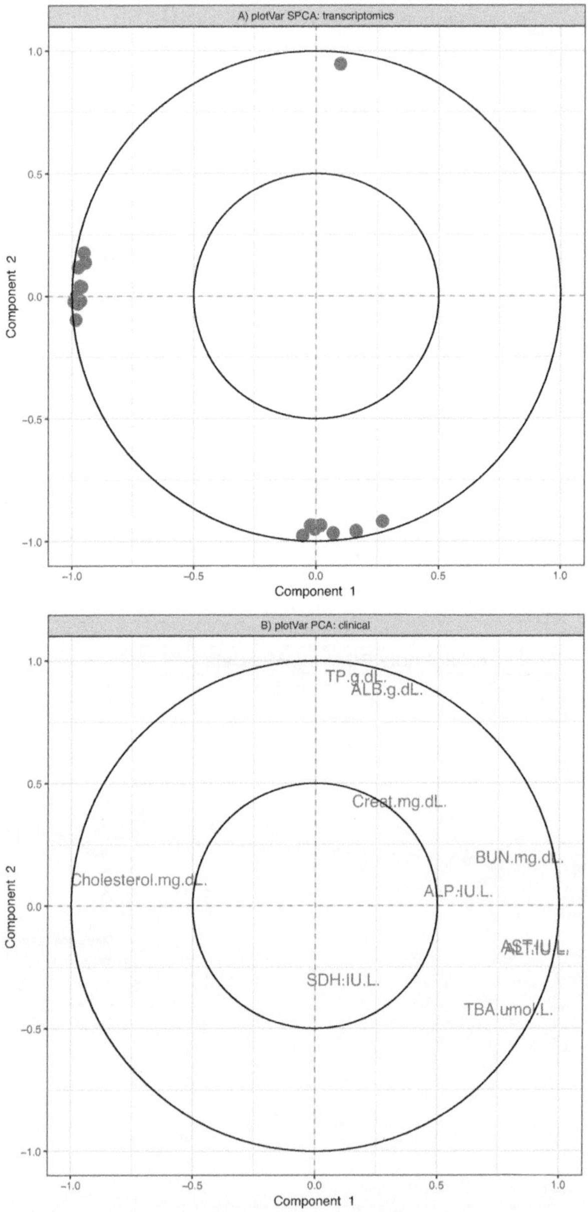

**Figure 7.12.** *Variable representation on each dataset analyzed separately when the data are projected on the first two principal components of (A) a sparse PCA for the transcriptomics data (gene names are not displayed for readbility reasons) and (B) PCA for the clinical data*

|  | Transcriptomics | | | | | | | | |
|---|---|---|---|---|---|---|---|---|---|
| PC | PC1 | PC2 | PC3 | PC4 | PC5 | PC6 | PC7 | PC8 | ... |
| Explained variance (%) | 28.1 | 14.9 | 8.0 | 5.4 | 4.6 | 3.9 | 3.4 | 2.5 | ... |
|  | Clinical | | | | | | | | |
| PC | PC1 | PC2 | PC3 | PC4 | PC5 | PC6 | PC7 | PC8 | ... |
| Explained variance (%) | 40.8 | 22.2 | 12.5 | 8.9 | 6.8 | 4.8 | 2.6 | 0.9 | ... |

**Table 7.2.** *Proportion of explained variance for the first eight principal components with respect to the total variance of the data. Top, for the transcriptomic dataset; bottom, for the clinical dataset*

### 7.6.3.2. Supervised analysis

We considered the outcome "time of exposure" that has four categories. Figure 7.13(A) shows that PLS-DA is able to separate the different times of exposure rather well. The separation is shown in Figure 7.13(B) with the sparse PLS-DA. To obtain this result, we chose to select 10 transcripts on each component. The number of transcripts to be selected has been chosen arbitrarily here for illustration purposes, but it can be determined using a cross-validation approach implemented in the tune.splsda function.

Instead of representing the correlation circle plot as shown with PCA, in Figure 7.14, we display a barplot of the loadings for the selected variables on each component, obtained using the plotLoadings function.

Selected variables are represented from bottom to top in increasing order of the absolute value of their loading. For instance, the gene with ID A_43_P13183 at the bottom has the highest loading in absolute value for the component 1 (Figure 7.14(A)). The color (orange) indicates the sample group for which this variable obtains the highest mean value. Consistently with the representation of samples, this gene seems to be overexpressed in the 18h group located on the left of Figure 7.13(B). The gene A_42_P508273 is ranked fourth in the barplot for component 1 and has a negative loading value. The barplot color (green) indicates that the gene is on average overexpressed in the 48h group. Once again, this is consistent with the location of the samples belonging to the 48h group in the representation of the individuals. It can also be noticed that selected genes with positive loading values are all overexpressed in the 6h and 24h groups, in agreement with the location of the individuals on the right of the representation of individuals. Regarding the barplot for component 2 (Figure 7.14(B)), the bars represent negative values and are mainly green, which is consistent with the location of the samples from the 48h group at the bottom of Figure 7.13(B).

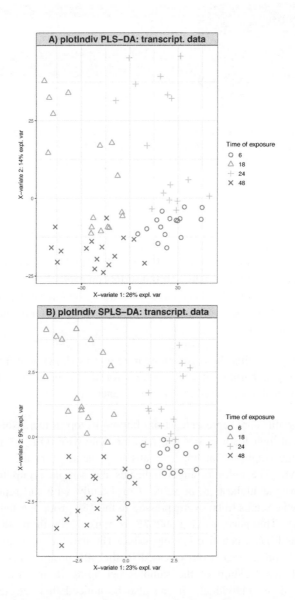

**Figure 7.13.** *Sample representation from a PLS-DA (A) and a sparse PLS-DA (B) on the transcriptomics data with the time of exposure considered as a categorical outcome. The sparse PLS-DA includes 10 selected transcripts (the number of genes was set arbitrarily for illustration purposes) on each component*

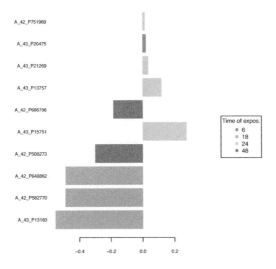

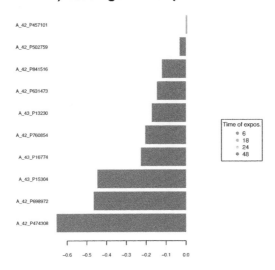

**Figure 7.14.** *Representation of the loadings on the first two components of a sparse PLS-DA: (A) component 1; (B) component 2*

Regarding the clinical dataset, the selection is not relevant as it only includes 10 variables. The results of a PLS-DA are presented in Figure 7.15. The discrimination of the four groups of exposure time seems to be more difficult on this dataset than the gene expression dataset. One reason can be a greater influence of the dose on the clinical variables. The dose factor has not been taken into account in the analyses presented so far.

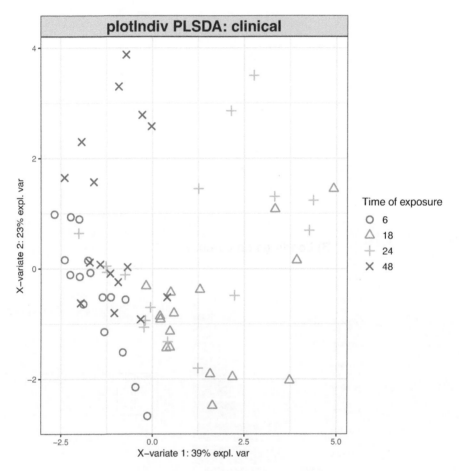

**Figure 7.15.** Sample representation from a PLS-DA on the clinical data where the outcome is the time of exposure

## 7.6.4. Integrative analysis

In section 7.6.3, we showed that unsupervised or supervised multivariate methods on each dataset considered independently could give some useful insight into the data. In this section, we illustrate an integrative analysis for two numerical datasets for both unsupervised and supervised frameworks.

### 7.6.4.1. Unsupervised analysis

We applied PLS to integrate the transcriptomics and clinical data. The representation of individuals is represented in Figure 7.16(A). It mainly highlights similar behaviors to those obtained with PCA (Figure 7.11(A)). For instance, low-dose samples (represented as triangles) are better clustered than high-dose samples. The first component mainly separate samples with high dose at time 18 h and 24 h on the left (orange and gray circles). The samples with high dose at 48 h located at the top of the plot are highlighted with the second component.

We also applied a sPLS that includes all 10 clinical variables, while we optimally selected 10 transcripts on each component. Our choice was motivated by a preliminary stability analysis to ensure that all selected features were repeatedly selected across the different partitions of the data. The representation of individuals on the first two components (Figure 7.16(B)) shows that samples at low doses (triangles) gather together on the right-hand side of the plot. Samples at high doses at 6 h of exposure (blue circles) are close to the low-dose samples. Other samples are more scattered, but still present some patterns that can be identified and interpreted. Indeed, we can consider that regardless of the time of exposure, the low dose has a small effect in both transcriptomics and clinical data, and samples cluster together. With a high dose, 6 h of exposure time does not create large changes in variation in the transcriptomics and clinical data, which may explain the location of these samples close to the low-dose samples. However, with a higher exposure time, the effects are stronger and generate much more variability for both clinical measurements and genes expression. A gradient can be identified from bottom to top with increasing time of exposure (18 h in orange, 24 h in gray and 48 h in green circle symbols). This figure demonstrates the benefit of integrating the two datasets compared to analyzing each dataset individually, as it highlighted some biological variation of interest.

To complete the overview of this unsupervised integrative approach, we look at the representation of variables for the sPLS. Two plots are provided in Figure 7.17. On the left-hand side (Figure 7.17(A)), the correlation circle plot represents all clinical variables with their labels and the selected transcripts (only symbols are displayed because of the overlapping of the labels otherwise). On the right-hand side (Figure 7.17(B)), loading plots are provided for the first component. Two barplots are displayed corresponding to each set of variables.

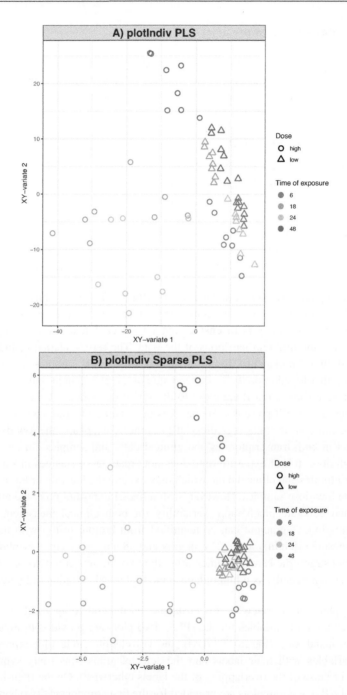

**Figure 7.16.** *Representation of individuals after a PLS (A) and a sparse PLS (B) selecting 10 variables on each component on the transcriptomics and clinical data*

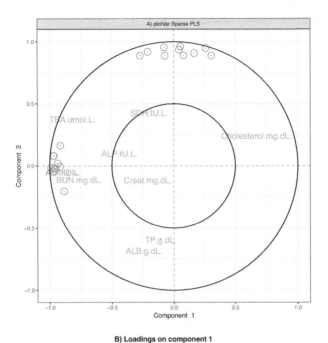
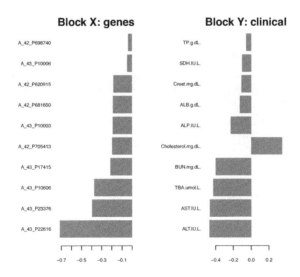

**Figure 7.17.** *Representation of the variables selected by SPLS: (A) in the correlation circle plot and (B) through the loadings for the first component*

We note that all clinical variables, with the exception of Cholesterol, have negative coordinates (left-hand side of the plot). This indicates that samples with high doses have the highest values for clinical variables, specifically for ALT, AST, TBA and BUN. Dots on the left-hand side of the plot show that transcripts are positively correlated to these clinical variables. They can be further identified on the loading plot. For instance, A_43_P22616 is the gene with the highest (negative) loading value on the first sPLS component and is thus associated with high values of ALT, AST, TBA and BUN in high-dose samples. Such interpretations represent the premise of more thorough biological investigations.

### 7.6.4.2. Supervised analysis

In this section, we illustrate the method that can be viewed as the ultimate goal of this study: a sparse multi-block supervised analysis. It consists of simultaneously analyzing the two numerical datasets, as well as a categorical outcome. Thus, we consider the integration of three data matrices: the transcriptomics and the clinical datasets, as well as the time exposure outcome coded as a dummy matrix internally in the method. Note that while all datasets have been analyzed in an integrative way, we obtain a set of components for each dataset, which enables us to carefully investigate each dataset. In this context, we also investigate the influence of the design matrix comparing the two designs presented in Table 7.3. The first one, referred to as full design, relates the two numerical datasets and the outcome in an equivalent manner. The second design, referred to as DA-oriented design, favors discrimination by assigning a small coefficient (0.1) between transcriptomics and clinical datasets. For both analyses, we focus on the sparse version to select 10 transcripts on each component, while all clinical variables are kept in the analysis.

| Full design | Transcriptomics | Clinical | Outcome |
|---|---|---|---|
| Transcriptomics | 0 | 1 | 1 |
| Clinical | 1 | 0 | 1 |
| Outcome | 1 | 1 | 0 |
| DA-oriented design | Transcriptomics | Clinical | Outcome |
| Transcriptomics | 0 | 0.1 | 1 |
| Clinical | 0.1 | 0 | 1 |
| Outcome | 1 | 1 | 0 |

**Table 7.3.** *Design matrix for multi-block analysis. A value of 1 indicates that the covariance between the two datasets is maximized in the model*

The results of the multi-block sparse PLS-DA are represented in Figures 7.18 and 7.20.

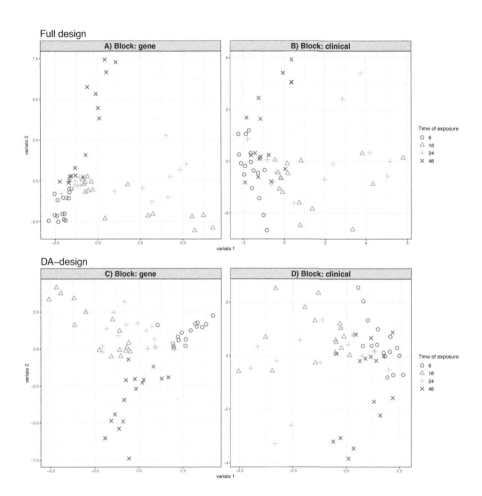

**Figure 7.18.** *Sample representation after a multi-block sparse PLS-DA when the transcriptomics and clinical data are integrated with the outcome* Time of exposure *on the subspace spanned by the transcriptomics data (A–C) and by the clinical data (B–D). Panels (A) and (B) correspond to the full design, and panels (C) and (D) correspond to the DA-oriented design*

The sample plots show that the discrimination is superior on the bottom left plot. The DA-oriented design maximizes the discrimination of the categorical outcome, rather than the integration of the two numerical datasets. Note that the discrimination is not perfect, as the samples are to be located on a three-pointed star. This is consistent with the dose effect that we observed in the previous analyses where we also disregarded the dose information. Indeed, as shown in Figure 7.19 representing the samples color coded according to dose intake, samples with low doses are located in the center of the plot, while samples with high doses scatter in different directions (upper right corner for 6h: blue circles at the bottom, for 48h: green crosses).

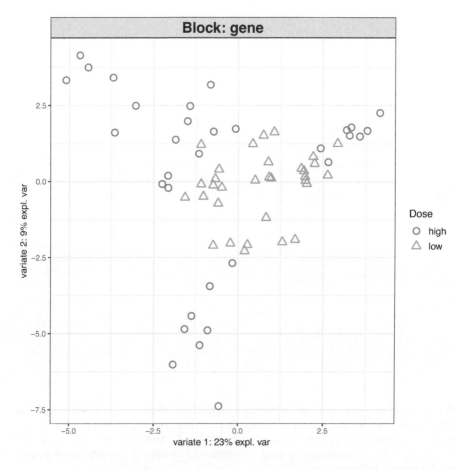

**Figure 7.19.** *Individuals' plots similar to Figure 7.18(C), but with individuals color-coded according to dose*

We interpret the results of this multi-blocks sparse discriminant analysis using clustered image maps to visualize the variables in Figure 7.20. Other graphical outputs such as relevance networks, correlation circle plots and loading plots can be considered.

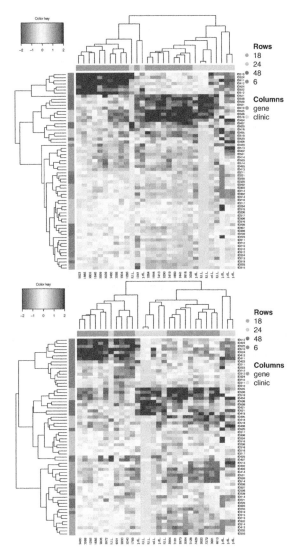

**Figure 7.20.** *Clustered image maps representing the results of a sparse multi-block PLS-DA with the time of exposure as the categorical outcome with (A) the full design and (B) the DA-oriented design*

These graphical representations confirm that the discrimination of the samples is better achieved with a DA-oriented design. The samples time of exposure are color coded on the left column of the plot. In Figure 7.20(B), we observe three types of sample clusters from top to bottom: the 6 h samples (blue), a mixture of 24 h and 48 h samples (gray and green), and a mixture of 18 h and 48 h (orange and green). We do not observe such distinction using the full design (Figure 7.20(A)).

The first row indicates the type of variables (clinical in yellow, transcripts in pink) selected or modeled by the method and their relationship with respect to the clusters of samples identified. Complemented with other types of graphical outputs and downstream statistical analyses, the integrative analysis opens new avenues for further biological interpretations and validations.

## 7.7. Conclusion

Integrating multiple 'omics matching datasets is without a doubt bringing more insight into a biological system than analyzing each dataset individually. Such analysis has the enormous potential of bringing new biological knowledge and discoveries. Multivariate projection-based methods are promising approaches to achieve such objective. In this chapter, we presented a wide range of approaches to answer different types of biological questions. We illustrated their application on a rat liver toxicity study, ranging from the exploration of one single dataset, to the search of biomarkers in one or several 'omics and/or clinical datasets. By leveraging on sample and variable graphical representations, we compared the different types of analyses. While much progress has been made in the field of 'omics analysis, more statistical developments are still needed to fully address the multiple 'omics data integration gap, in particular to accommodate for more complex experimental designs, such as longitudinal studies and the combination of independent studies.

## 7.8. Acknowledgments

The authors would like to thank the International Centre for Mathematics and Computer Science (CIMI) in Toulouse, which funded a part of this research. KALC was supported in part by the National Health and Medical Research Council Career Development Fellowship (GNT1159458).

## 7.9. Appendix: reproducible R code

### 7.9.1. *Toy examples*

```
library(mixOmics)   # Version 6.3.1
library(gridExtra)  # Version 2.2.1
library(GGally)     # Version 1.4.0

set.seed(1)

## PCA

### 3 uncorrelated variables
pca.data1 <- as.data.frame(matrix(runif(150),nc=3))
round(cor(pca.data1),2)
pca.res.data1 <- pca(pca.data1, ncomp=3)
summary(pca.res.data1)

### 2 correlated variables + 1
pca.data2 <- as.data.frame(matrix(rnorm(150),nc=3))
pca.data2[,2] = pca.data2[,1]+rnorm(50,0,0.5)
round(cor(pca.data2),2)
pca.res.data2 <- pca(pca.data2, ncomp=3)
summary(pca.res.data2)

### 3 correlated variables
pca.data3 <- as.data.frame(matrix(rnorm(150),nc=3))
pca.data3[,2] = pca.data3[,1]+rnorm(50,0,0.5)
pca.data3[,3] = pca.data3[,1]+rnorm(50,0,0.5)
round(cor(pca.data3),2)
pca.res.data3 <- pca(pca.data3, ncomp=3)
summary(pca.res.data3)

pca.data1.pairs <- ggpairs(pca.data1)
pca.data2.pairs <- ggpairs(pca.data2)
pca.data3.pairs <- ggpairs(pca.data3)

pca.data1.pairs
pca.data2.pairs
pca.data3.pairs

## PLS-DA

plsda.data <- matrix(rep(0,150),nc=3)
colnames(plsda.data) <- c("V1","V2","V3")
plsda.data[,1] <- rnorm(50, mean = 0, sd = 10)
plsda.data[,2] <- rnorm(50, mean = 0, sd = 5)
```

```
plsda.data[,3] <- c(rnorm(25, mean=-1, sd=1),
                    rnorm(25, mean=1, sd=1))
Group <- rep(c("A","B"), each=25)
plsda.data.df <- data.frame(plsda.data, Group=Group)
summary(plsda.data.df)

plsda.data.V1.boxplot <- ggplot(plsda.data.df) +
  scale_y_continuous(limits=c(-25,25)) +
  geom_boxplot(aes(x=Group, y=V1)) +
  geom_point(aes(x=Group, y=V1), alpha=0.1,
             position = position_jitter(w=.1))

plsda.data.V2.boxplot <- ggplot(plsda.data.df) +
  scale_y_continuous(limits=c(-25,25)) +
  geom_boxplot(aes(x=Group, y=V2)) +
  geom_point(aes(x=Group, y=V2), alpha=0.1,
             position = position_jitter(w=.1))

plsda.data.V3.boxplot <- ggplot(plsda.data.df) +
  scale_y_continuous(limits=c(-25,25)) +
  geom_boxplot(aes(x=Group, y=V3)) +
  geom_point(aes(x=Group, y=V3), alpha=0.1,
             position = position_jitter(w=.1))

grid.arrange(plsda.data.V1.boxplot,
             plsda.data.V2.boxplot,
             plsda.data.V3.boxplot, ncol=3, nrow=1)

pca.res.plsda.data <- pca(plsda.data, ncomp=3)
summary(pca.res.plsda.data)
round(pca.res.plsda.data$loadings$X,3)

## PLS

pls.data.X <- matrix(rnorm(250), ncol=5)
colnames(pls.data.X) <- paste0("X",1:5)
pls.data.Y <- matrix(rep(0, 150), ncol=3)
colnames(pls.data.Y) <- paste0("Y",1:3)
pls.data.Y[,1] <- pls.data.X[,1] + rnorm(50,0,0.5)
pls.data.Y[,2] <- rnorm(50)
pls.data.Y[,3] <- -2*pls.data.X[,2] + 2*pls.data.X[,3] +
                   rnorm(50,0,0.5)
pls.data.df <- data.frame(pls.data.X, pls.data.Y)

ggcorr(pls.data.df, palette = "RdBu", label = TRUE,
       geom = "tile")
```

```
pls.res.data.pls <- pls(pls.data.X, pls.data.Y)
plotVar(pls.res.data.pls)

## DIABLO

block.pls.data.X <- matrix(rnorm(250), ncol=5)
colnames(block.pls.data.X) <- paste0("X",1:5)

block.pls.data.Y <- matrix(rep(0, 150), ncol=3)
colnames(block.pls.data.Y) <- paste0("Y",1:3)

block.pls.data.Y[,1] <- block.pls.data.X[,1] +
                        rnorm(50,0,0.5)
block.pls.data.Y[,2] <- rnorm(50)
block.pls.data.Y[,3] <- -2*block.pls.data.X[,2] +
  2*block.pls.data.X[,3] + rnorm(50,0,0.5)

block.pls.data.Z <- matrix(rep(0, 400), ncol=8)
colnames(block.pls.data.Z) <- paste0("Z",1:8)

block.pls.data.Z[,1:6] <- rnorm(300)
block.pls.data.Z[,7] <- c(rnorm(25, mean=-1, sd=1),
                  rnorm(25, mean=1, sd=1))
block.pls.data.Z[,8] <- -2*block.pls.data.X[,2] +
  2*block.pls.data.X[,3] + rnorm(50,0,0.5)

Group <- rep(c("A","B"), each=25)

block.pls.data.df <- data.frame(block.pls.data.X,
                      block.pls.data.Y,
                      block.pls.data.Z,
                      Group)

ggcorr(block.pls.data.df[,-17], palette = "RdBu",
  label = TRUE, geom = "tile")

block.pls.data.Z1.boxplot <- ggplot(block.pls.data.df) +
  scale_y_continuous(limits=c(-5,5)) +
  geom_boxplot(aes(x=Group, y=Z1)) +
  geom_point(aes(x=Group, y=Z1), alpha=0.1,
             position = position_jitter(w=.1))

block.pls.data.Z2.boxplot <- ggplot(block.pls.data.df) +
  scale_y_continuous(limits=c(-5,5)) +
  geom_boxplot(aes(x=Group, y=Z2)) +
  geom_point(aes(x=Group, y=Z2), alpha=0.1,
             position = position_jitter(w=.1))
```

```
block.pls.data.Z3.boxplot <- ggplot(block.pls.data.df) +
  scale_y_continuous(limits=c(-5,5)) +
  geom_boxplot(aes(x=Group, y=Z3)) +
  geom_point(aes(x=Group, y=Z3), alpha=0.1,
             position = position_jitter(w=.1))

block.pls.data.Z4.boxplot <- ggplot(block.pls.data.df) +
  scale_y_continuous(limits=c(-5,5)) +
  geom_boxplot(aes(x=Group, y=Z4)) +
  geom_point(aes(x=Group, y=Z4), alpha=0.1,
             position = position_jitter(w=.1))

block.pls.data.Z5.boxplot <- ggplot(block.pls.data.df) +
  scale_y_continuous(limits=c(-5,5)) +
  geom_boxplot(aes(x=Group, y=Z5)) +
  geom_point(aes(x=Group, y=Z5), alpha=0.1,
             position = position_jitter(w=.1))

block.pls.data.Z6.boxplot <- ggplot(block.pls.data.df) +
  scale_y_continuous(limits=c(-5,5)) +
  geom_boxplot(aes(x=Group, y=Z6)) +
  geom_point(aes(x=Group, y=Z6), alpha=0.1,
             position = position_jitter(w=.1))

block.pls.data.Z7.boxplot <- ggplot(block.pls.data.df) +
  scale_y_continuous(limits=c(-5,5)) +
  geom_boxplot(aes(x=Group, y=Z7)) +
  geom_point(aes(x=Group, y=Z7), alpha=0.1,
             position = position_jitter(w=.1))

block.pls.data.Z8.boxplot <- ggplot(block.pls.data.df) +
  scale_y_continuous(limits=c(-5,5)) +
  geom_boxplot(aes(x=Group, y=Z8)) +
  geom_point(aes(x=Group, y=Z8), alpha=0.1,
             position = position_jitter(w=.1))

grid.arrange(block.pls.data.Z1.boxplot,
             block.pls.data.Z2.boxplot,
             block.pls.data.Z3.boxplot,
             block.pls.data.Z4.boxplot,
             block.pls.data.Z5.boxplot,
             block.pls.data.Z6.boxplot,
             block.pls.data.Z7.boxplot,
             block.pls.data.Z8.boxplot,
             ncol=4, nrow=2)
```

```
blocks <- list(X=block.pls.data.df[,1:5],
               Y=block.pls.data.df[,6:8],
               Z=block.pls.data.df[,9:16])

res.block.plsda <- block.plsda(blocks, Group)

plotIndiv(res.block.plsda, group=Group,
          ind.names=FALSE, legend = TRUE)

plotVar(res.block.plsda)
```

### 7.9.2. *Liver toxicity*

```
data(liver.toxicity)
help(liver.toxicity)

liver.toxicity$treatment$Dose.Group <-
   as.factor(c(rep("low",32), rep("high",32)))


## PCA

data.lt.gene <- liver.toxicity$gene

res.pca.lt.gene <- pca(data.lt.gene, scale=TRUE, ncomp=10)

plotIndiv(res.pca.lt.gene, ind.names = FALSE,
          title = "A) plotIndiv PCA: transcriptomics data",
          group=liver.toxicity$treatment$Time.Group,
          pch = as.numeric(factor(liver.toxicity$treatment
             $Dose.Group)),
          pch.levels =liver.toxicity$treatment$Dose.Group,
          legend = TRUE, legend.title = c("Time of
             exposure"),
          legend.title.pch = "Dose")

plotVar(res.pca.lt.gene, var.names = FALSE)

data.lt.clinic <- liver.toxicity$clinic
res.pca.lt.clinic <- pca(data.lt.clinic, scale=TRUE,
          ncomp=10)

plotIndiv(res.pca.lt.clinic, ind.names = FALSE,
          title = "B) plotIndiv PCA: clinical data",
          group=liver.toxicity$treatment$Time.Group,
          pch = as.numeric(factor(liver.toxicity$treatment
             $Dose.Group)),
```

```
                pch.levels =liver.toxicity$treatment$Dose.Group,
                legend = TRUE, legend.title = c("Time of
                    exposure"),
                legend.title.pch = "Dose")

plotVar(res.pca.lt.clinic,
        title = "plotVar PCA: clinical")

## Sparse PCA

res.spca.lt.gene <- spca(data.lt.gene, scale=TRUE,
                         ncomp=3, keepX=c(10,10,10))

plotIndiv(res.spca.lt.gene, ind.names = FALSE,
          group=liver.toxicity$treatment$Time.Group,
          pch = as.numeric(factor(liver.toxicity$treatment
              $Dose.Group)),
          pch.levels =liver.toxicity$treatment$Dose.Group,
          legend = TRUE)

plotVar(res.spca.lt.gene, var.names = FALSE, pch=16,
        title = "A) plotVar SPCA: transcriptomics")




## PLS-DA

res.plsda.lt.gene.time <- plsda(data.lt.gene,
                                liver.toxicity$treatment
                                  $Time.Group)

plotIndiv(res.plsda.lt.gene.time, ind.names = FALSE,
          title = "A) plotIndiv PLS-DA: transcript. data",
          legend = TRUE, legend.title = c("Time of exposure"))

res.plsda.lt.clinic.time <- plsda(data.lt.clinic,
                                  liver.toxicity$treatmen
                                    t$Time.Group)

plotIndiv(res.plsda.lt.clinic.time, ind.names = FALSE,
          legend = TRUE, legend.title = "Time of exposure",
          title = "plotIndiv PLSDA: clinical")

## Sparse PLS-DA

res.splsda.lt.gene.time <- splsda(data.lt.gene,
```

```
                              liver.toxicity$treatment
                                 $Time.Group,
                                 ncomp = 3, keepX =
                                          c(10,10,10))

plotIndiv(res.splsda.lt.gene.time, ind.names = FALSE,
          title = "B) plotIndiv SPLS-DA: transcript. data",
          legend = TRUE, legend.title = c("Time of
             exposure"))


plotVar(res.splsda.lt.gene.time)

plotLoadings(res.splsda.lt.gene.time, comp=1,
             contrib = "max",
             title = "A) Loadings on comp. 1",
             legend.title = "Time of expos.")

plotLoadings(res.splsda.lt.gene.time, comp=2,
             contrib = "max",
             title = "B) Loadings on comp. 2",
             legend.title = "Time of expos.")
## PLS
res.pls.lt.gene.clinic <- pls(data.lt.gene,data.lt.clinic)

plotIndiv(res.pls.lt.gene.clinic,rep.space="XY-variate",
          title = "A) plotIndiv PLS",
          ind.names=FALSE,
          group=liver.toxicity$treatment$Time.Group,
          pch = as.numeric(factor(liver.toxicity$treatment
             $Dose.Group)),
          pch.levels =liver.toxicity$treatment$Dose.Group,
          legend = TRUE, legend.title = c("Time of
             exposure"),
          legend.title.pch = "Dose")

plotVar(res.pls.lt.gene.clinic, var.names = c(FALSE, TRUE))

## Sparse PLS

res.spls.lt.gene.clinic <- spls(data.lt.gene,data.lt.clinic,
                                ncomp=3, keepX = c(10,10,10))

plotIndiv(res.spls.lt.gene.clinic,rep.space="XY-variate",
          title = "B) plotIndiv Sparse PLS",
          ind.names=FALSE,
```

```
                group=liver.toxicity$treatment$Time.Group,
                pch = as.numeric(factor(liver.toxicity$treatment
                   $Dose.Group)),
                pch.levels =liver.toxicity$treatment$Dose.Group,
                legend = TRUE, legend.title = c("Time of
                   exposure"),
                legend.title.pch = "Dose")

plotVar(res.spls.lt.gene.clinic, var.names = c(FALSE, TRUE))

plotLoadings(res.spls.lt.gene.clinic, comp=1, size.title = 1,
             title = "B) Loadings on component 1",
             subtitle=c("Block X: genes","Block Y:
                 clinical"))

## Multi-block

res.block.plsda.lt.gene.clinin.time <- block.plsda(
  X = list(gene = data.lt.gene,
           clinic = data.lt.clinic),
  Y = liver.toxicity$treatment$Time.Group)

plotIndiv(res.block.plsda.lt.gene.clinin.time)
plotVar(res.block.plsda.lt.gene.clinin.time)



## Sparse Multi-block

list.keepX <- list(gene = rep(10, 2), clinic = rep(10,2))

res.block.splsda.lt.gene.clinin.time <- block.splsda(
  X = list(gene = data.lt.gene,
           clinic = data.lt.clinic),
  Y = liver.toxicity$treatment$Time.Group,
  keepX = list.keepX)

plotIndiv(res.block.splsda.lt.gene.clinin.time,
          title = "Full design",
          subtitle = c("A) Block: gene","B) Block:
              clinical"),
          group=liver.toxicity$treatment$Time.Group,
          ind.names=FALSE, legend = TRUE,
          legend.title = "Time of exposure")

plotVar(res.block.splsda.lt.gene.clinin.time)

cimDiablo(res.block.splsda.lt.gene.clinin.time,
```

```
            color.blocks = color.mixo(5:6),
            color.Y=color.mixo(c(2,3,4,1)))

# Change design

design.da <- matrix(c(0,0.1,1,0.1,0,1,1,1,0), ncol=3)

res.block.splsda.lt.design.gene.clinic.time <- block.splsda(
  X = list(gene = data.lt.gene,
           clinic = data.lt.clinic),
  Y = liver.toxicity$treatment$Time.Group,
  keepX = list.keepX, design=design.da)

plotIndiv(res.block.splsda.lt.design.gene.clinic.time,
          title = "DA-design",
          subtitle = c("C) Block: gene","D) Block:
                       clinical"),
          group=liver.toxicity$treatment$Time.Group,
          ind.names=FALSE, legend = TRUE,
          legend.title = "Time of exposure")

plotIndiv(res.block.splsda.lt.design.gene.clinic.time,
          group=liver.toxicity$treatment$Dose.Group,
          ind.names=FALSE, legend = TRUE, blocks = 1,
          legend.title = "Dose")

cimDiablo(res.block.splsda.lt.design.gene.clinic.time,
          color.blocks = color.mixo(5:6),
          color.Y=color.mixo(c(2,3,4,1)))
```

## 7.10. References

Alter, O., Brown, P.O., Botstein, D. (2000). Singular value decomposition for genome-wide expression data processing and modeling. *Proceedings of the National Academy of Sciences*, 97(18), 10101–10106.

Ambroise, C. and McLachlan, G.J. (2002). Selection bias in gene extraction in tumour classification on basis of microarray gene expression data. *Proc. Natl. Acad. Sci. USA*, 99(1), 6562–6566.

Barker, M. and Rayens, W. (2003). Partial least squares for discrimination. *Journal of Chemometrics*, 17(3), 166–173.

Boulesteix, A. and Strimmer, K. (2007). Partial least squares: A versatile tool for the analysis of high-dimensional genomic data. *Briefings in Bioinformatics*, 8(1), 32.

Bushel, P.R., Wolfinger, R.D., Gibson, G. (2007). Simultaneous clustering of gene expression data with clinical chemistry and pathological evaluations reveals phenotypic prototypes. *BMC Systems Biology*, 1(1), 15.

Carninci, P., Kasukawa, T., Katayama, S., Gough, J., Frith, M., Maeda, N., Oyama, R., Ravasi, T., Lenhard, B., Wells, C. et al. (2005). The transcriptional landscape of the mammalian genome. *Science*, 309, 1559–1563.

Chun, H. and Keles, S. (2010). Sparse partial least squares regression for simultaneous dimension reduction and variable selection. *Journal of the Royal Statistical Society, Series B: Statistical Methodology*, 72(1), 3–25.

Chung, D. and Keles, S. (2010). Sparse partial least squares classification for high dimensional data. *Statistical Applications in Genetics and Molecular Biology*, 9(1), 17.

Combes, S., González, I., Déjean, S., Baccini, A., Jehl, N., Juin, H., Cauquil, L., Gabinaud, B., Lebas, F., Larzul, C. (2008). Relationships between sensory and physicochemical measurements in meat of rabbit from three different breeding systems using canonical correlation analysis. *Meat Science*, 80(3), 835–841.

Comon, P. (1992). Independent component analysis. In *Higher-order Statistics*, Lacoume, J.L. (ed.). Elsevier, Amsterdam.

Esposito Vinzi, V., Chin, W., Henseler, J., Wang, H. (eds) (2010). *Handbook of Partial Least Squares*. Springer, Berlin, Heidelberg.

Goble, C. and Stevens, R. (2008). State of the nation in data integration for bioinformatics. *Journal of Biomedical Informatics*, 41(5), 687–693.

Gomez-Cabrero, D., Abugessaisa, I., Maier, D., Teschendorff, A., Merkenschlager, M., Gisel, A., Ballestar, E., Bongcam-Rudloff, E., Conesa, A., Tegner, J. (2014). Data integration in the era of omics: Current and future challenges. *BMC Systems Biology*, 8 [Online]. Available at: https://bmcsystbiol.biomedcentral.com/articles/10.1186/1752-0509-8-S2-I1.

González, I., Déjean, S., Martin, P.G., Baccini, A. (2008). CCA: An R package to extend canonical correlation analysis. *Journal of Statistical Software*, 23(12), 1–14.

González, I., Déjean, S., Martin, P.G., Gonçalves, O., Besse, P., Baccini, A. (2009). Highlighting relationships between heterogeneous biological data through graphical displays based on regularized canonical correlation analysis. *Journal of Biological Systems*, 17(2), 173–199.

González, I., Lê Cao, K.-A., Davis, M., Déjean, S. (2012). Visualising associations between paired "omics" data sets. *BioData Mining*, 5(1), 19.

Günther, O.P., Shin, H., Ng, R.T., McMaster, W.R., McManus, B.M., Keown, P.A., Tebbutt, S.J., Lê Cao, K.-A. (2014). Novel multivariate methods for integration of genomics and proteomics data: Applications in a kidney transplant rejection study. *Omics: A Journal of Integrative Biology*, 18(11), 682–695.

Helwig, N. (2018). ica: Independent Component Analysis. R package version 1.0-2.

Hotelling, H. (1936). Relations between two sets of variables. *Biometrika*, 28(3–4), 321–377.

Hyvärinen, A. and Oja, E. (2000). Independent component analysis: Algorithms and applications. *Neural Networks*, 13(4), 411–430.

Joyce, A. and Palsson, B. (2006). The model organism as a system: Integrating "omics" data sets. *Nature Reviews. Molecular Cell Biology*, 7, 198–210.

Lê Cao, K.-A., Rossouw, D., Robert-Granié, C., Besse, P. (2008). A sparse PLS for variable selection when integrating omics data. *Statistical Applications in Genetics and Molecular Biology*, 7, 35.

Lê Cao, K.-A., Martin, P.G., Robert-Granié, C., Besse, P. (2009). Sparse canonical methods for biological data integration: Application to a cross-platform study. *BMC Bioinformatics*, 10(1), 34.

Lê Cao, K.-A., Boitard, S., Besse, P. (2011). Sparse PLS discriminant analysis: Biologically relevant feature selection and graphical displays for multiclass problems. *BMC Bioinformatics*, 12(1), 253.

Lee, A., Shannon, C., Amenyogbe, N., Bennike, T., Diray-Arce, J., Idoko, O., Gill, E., Ben-Othman, R., Pomat, W., Haren, S. et al. (2019). Dynamic molecular changes during the first week of human life follow a robust developmental trajectory. *Nature Communications*, 10 [Online]. Available at: https://www.nature.com/articles/s41467-019-08794-x/#citeas.

Liquet, B., Lê Cao, K.-A., Hocini, H., Thiébaut, R. (2012). A novel approach for biomarker selection and the integration of repeated measures experiments from two assays. *BMC Bioinformatics*, 13(1), 325.

Marchini, J., Heaton, C., Ripley, B. (2019). fastICA: FastICA algorithms to perform ICA and projection pursuit. R package version 1.2-2.

McLendon, R., Friedman, A., Bigner, D., Van Meir, E., Brat, D., Mastrogianakis, G., Olson, J., Mikkelsen, T., Lehman, N., Aldape, K. et al. (2008). Comprehensive genomic characterization defines human glioblastoma genes and core pathways. *Nature*, 455, 1061–1068 [Online]. Available at: https://www.nature.com/articles/nature07385#citeas.

de Noord, O.E. and Theobald, E.H. (2005). Multilevel component analysis and multilevel PLS of chemical process data. *Journal of Chemometrics*, 19(5–7), 301–307.

Parkhomenko, E., Tritchler, D., Beyene, J. (2009). Sparse canonical correlation analysis with application to genomic data integration. *Statistical Applications in Genetics and Molecular Biology*, 8(1).

Rohart, F., Gautier, B., Singh, A., Lê Cao, K.-A. (2017a). mixOmics: An R package for 'omics feature selection and multiple data integration. *PLoS Computational Biology*, 13(11) [Online]. Available at: https://journals.plos.org/ploscompbiol/article?id=10.1371/journal.pcbi.1005752.

Rohart, F., Matigian, N., Eslami, A., Bougeard, S., Lê Cao, K.-A. (2017b). MINT: A multivariate integrative method to identify reproducible molecular signatures across independent experiments and platforms. *BMC Bioinformatics*, 18(1), 128.

Schäfer, J. and Strimmer, K. (2005). A shrinkage approach to large-scale covariance matrix estimation and implications for functional genomics. *Statistical Applications in Genetics and Molecular Biology*, 4(1).

Schloerke, B., Crowley, J., Cook, D., Briatte, F., Marbach, M., Thoen, E., Elberg, A., Larmarange, J. (2018). GGally: Extension to "ggplot2". R package version 1.4.0.

Scholz, M., Gatzek, S., Sterling, A., Fiehn, O., Selbig, J. (2004). Metabolite fingerprinting: Detecting biological features by independent component analysis. *Bioinformatics*, 20(15), 2447–2454.

Singh, A., Shannon, C., Gautier, B., Rohart, F., Vacher, M., Tebbutt, S., Lê Cao, K.-A. (2019). DIABLO: An integrative approach for identifying key molecular drivers from multi-omics assays. *Bioinformatics*, 35(17), 3055–3062 [Online]. Available at: https://academic.oup.com/bioinformatics/article/35/17/3055/5292387.

Tenenhaus, M. and Hanafi, M. (2010). *A Bridge Between PLS Path Modeling and Multi-Block Data Analysis*. Springer, Berlin, Heidelberg.

Tenenhaus, A. and Tenenhaus, M. (2011). Regularized generalized canonical correlation analysis. *Psychometrika*, 76(2), 257–284.

Tenenhaus, A., Philippe, C., Guillemot, V., Lê Cao, K.-A., Grill, J., Frouin, V. (2014). Variable selection for generalized canonical correlation analysis. *Biostatistics*, 15(3), 569–583.

The ENCODE Project Consortium (2004). The ENCODE (ENCyclopedia Of DNA Elements) Project. *Science*, 306(5696), 636–640 [Online]. Available at: https://www.science.org/doi/10.1126/science.1105136.

Tibshirani, R. (1996). Regression shrinkage and selection via the lasso. *Journal of the Royal Statistical Society, Series B: Methodological*, 58(1), 267–288.

van Velzen, E.J.J., Westerhuis, J.A., van Duynhoven, J.P.M., van Dorsten, F.A., Hoefsloot, H.C.J., Jacobs, D.M., Smit, S., Draijer, R., Kroner, C.I., Smilde, A.K. (2008). Multilevel data analysis of a crossover designed human nutritional intervention study. *Journal of Proteome Research*, 7, 4483–4491.

Waaijenborg, S., De Witt Hamer, P., Zwinderman, A. (2008). Quantifying the association between gene expressions and DNA-markers by penalized canonical correlation analysis. *Statistical Applications in Genetics and Molecular Biology*, 7(3).

Westerhuis, J.A., van Velzen, E.J.J., Hoefsloot, H.C.J., Smilde, A.K. (2010). Multivariate paired data analysis: Multilevel PLSDA versus OPLSDA. *Metabolomics*, 6(1), 119–128.

Wienkoop, S., Morgenthal, K., Wolschin, F., Scholz, M., Selbig, J., Weckwerth, W. (2008). Integration of metabolomic and proteomic phenotypes: Analysis of data covariance dissects starch and RFO metabolism from low and high temperature compensation response in Arabidopsis thaliana. *Molecular & Cellular Proteomics: MCP*, 7, 1725–1736.

Witten, D. and Tibshirani, R. (2009). Extensions of sparse canonical correlation analysis with applications to genomic data. *Statistical Applications in Genetics and Molecular Biology*, 8(28).

Witten, D., Tibshirani, R., Hastie, T. (2009). A penalized matrix decomposition, with applications to sparse principal components and canonical correlation analysis. *Biostatistics*, 10, 515–534.

Wold, S., Sjöström, M., Eriksson, L. (2001). PLS-regression: A basic tool of chemometrics. *Chemometrics and Intelligent Laboratory Systems*, 58(2), 109–130.

Yao, F., Coquery, J., Lê Cao, K.-A. (2012). Independent principal component analysis for biologically meaningful dimension reduction of large biological data sets. *BMC Bioinformatics*, 13(1), 24.

Zhang, W., Li, F., Nie, L. (2009). Integrating multiple "omics" analysis for microbial biology: Application and methodologies. *Microbiology*, 156(pt 2), 287–301.

# List of Authors

Christophe AMBROISE
LaMME
Université Paris-Saclay
CNRS
Université d'Évry
Évry-Courcouronnes
France

Sarah COHEN-BOULAKIA
Université Paris-Saclay
CNRS
LISN
Orsay
France

Gwendal CUEFF
Université Paris-Saclay
INRAE
AgroParisTech
IJPB
Versailles
France

Olivier DAMERON
Université de Rennes
INRIA
CNRS
IRISA
France

Sébastien DÉJEAN
Institut de Mathématiques de Toulouse
CNRS
Université de Toulouse
France

Christine FROIDEVAUX
Université Paris-Saclay
CNRS
LISN
Orsay
France

Florent GUINOT
Institut Roche
Boulogne-Billancourt
and
LaMME
Université Paris-Saclay
CNRS
Université d'Évry
Évry-Courcouronnes
France

Kim-Anh LÊ CAO
School of Mathematics and Statistics
Melbourne Integrative Genomics
Australia

Frédéric LEMOINE
Institut Pasteur
Université Paris Cité
France

Céline LÉVY-LEDUC
Université Paris-Saclay
AgroParisTech
INRAE
MIA Paris-Saclay
France

Jérôme MARIETTE
Université de Toulouse
INRAE
MIAT
Castanet-Tolosan
France

Marie-Laure MARTIN-MAGNIETTE
IPS2
Université Paris-Saclay
CNRS
INRAE
Université d'Évry
Université Paris Cité
Gif-sur-Yvette
and
MIA Paris-Saclay
Université Paris-Saclay
AgroParis Tech
INRAE
France

Marie PERROT-DOCKÈS
Université Paris Cité
MAP5
CNRS
France

Loïc RAJJOU
Université Paris-Saclay
INRAE
AgroParisTech
IJPB
Versailles
France

Bastien RANCE
Hôpital Européen Georges Pompidou
AP-HP
Paris
and
Centre de Recherche des Cordeliers
INSERM
Université Paris Cité
France

Guillem RIGAILL
IPS2
Université Paris-Saclay
CNRS
INRAE
Université d'Évry
Université Paris Cité
Gif-sur-Yvette
and
LaMME
Université Paris-Saclay
CNRS
Université d'Évry
Évry-Courcouronnes
France

Marie SZAFRANSKI
LaMME
Université Paris-Saclay
CNRS
Université d'Évry
Évry-Courcouronnes
and
ENSIIE
France

Nathalie VIALANEIX
Université de Toulouse
INRAE
MIAT
Castanet-Tolosan
France

Maxime WACK
Hôpital Européen Georges Pompidou
AP-HP
Paris
and
Centre de Recherche des Cordeliers
INSERM
Université Paris Cité
France

# Index

## A, C

ankylosing spondylitis (AS), 118, 142, 143, 146
class, 28–30, 34, 36, 37, 39–42, 45, 132, 134–136, 158–160, 169, 206–209, 211, 220, 221, 224
classification, 3, 5, 7, 16, 133–136, 138, 143, 154, 158, 159, 163, 167, 168, 170, 171, 201, 207–209
cohort, 4, 12, 13, 119, 123, 129, 142
complex/complexity, 4, 5, 7, 9, 10, 13, 15, 16, 18, 25–27, 29, 37, 40, 43, 46, 54–56, 61, 66, 67, 77–80, 100, 119, 121, 127, 131, 132, 134–136, 144, 158, 163, 166–168, 199, 215, 238
compression, 117, 132, 133, 137
Conda, 62, 66–68
connected graph, 29
correlation, 94, 96, 119, 126, 131, 132, 142, 146, 161, 163, 168, 173, 177, 178, 196–199, 203, 206, 210–222, 227, 231, 233, 234, 237, 239
circle, 213, 214, 219–221, 227, 231, 233, 237
covariance, 89, 92–95, 142, 162, 198, 199, 201, 207, 209, 211–213, 215, 234
criterion, 89, 97–99, 134, 159, 160, 162, 164, 170, 171, 173–175, 198, 199, 211, 212, 215, 217, 224
cut level, 133, 138–140, 143, 144, 146
CWL (Common Workflow Language), 60, 65, 66

## D, E

data
    deluge, 26
    new category of, 5
    science, 26, 46, 54
dereferencing, 32
design matrix, 215–217, 234
DNA chips, 120, 121
Docker, 62, 65–68, 75, 77
encapsulation, 60–62, 64, 68, 72, 80

endpoint, 36, 40–43, 45, 46
entity matching, 28, 45
execution environment, 55, 58, 62, 63, 65, 67, 68, 72

## G, H, I

Galaxy, 59, 61, 64–68, 72, 75, 76
GWAS (genome-wide association studies), 117–120, 125, 126, 132, 133, 138, 139, 142, 143, 146
haplotype, 127, 128, 138
hierarchical, 9, 132–138, 143, 159, 160, 222
HIS (hospital information system), 4, 5, 7, 11–13
identifier, 7, 9, 28, 31, 33, 43, 63
IRI (Internationalized Resource Identifier), 31, 32

## K, L, M

kernel, 130, 151–163, 165–181, 183–187
   trick, 155, 156, 158, 161
knowledge, 10, 14, 15, 27, 29–32, 36–40, 42–44, 46, 66, 139, 145, 146, 196, 216, 238
large dimension, 15, 211
Lasso (least absolute shrinkage and selection operator), 89, 92, 93, 96–99, 109, 111, 127, 143, 144, 200, 202, 211, 212
linear combination, 198–200, 207, 208, 211, 217, 224
marker, 118–121, 124–130, 133, 138, 146
missing heritability, 126
modularization, 60, 61

multiple, 5, 6, 12, 14, 27, 33, 39, 46, 55, 58–60, 72, 75, 125–127, 132, 133, 136, 138, 141, 143, 169, 171–173, 196, 197, 208, 220, 238

## N, O, P

nextflow, 59, 63–66, 68, 69, 72, 73, 75, 77
ontology, 7, 10, 17, 28, 30–32, 34, 36, 37, 39–42, 46, 60, 61, 65, 79
parsimony, 96–101, 161, 167, 173, 175, 178
phenotype, 17, 26, 43, 117–119, 123–126, 128, 129, 131, 133, 136–138, 141, 144–146, 151, 200, 211
polymorphism, 119–121, 124, 126, 142
principal component, 138, 157, 161–163, 167, 197, 198, 201, 203, 204, 225–227
provenance, 63, 64, 67–69, 79–81

## R, S, T

reasoning, 9, 18, 30, 31, 37, 40–42, 46
regression, 127, 130–132, 138–141, 143, 144, 146, 167, 197, 211
   penalized, 127, 138–140
reproducibility, 19, 21, 55, 56, 80
research, 3–5, 7, 10, 12–17, 19, 21, 26, 28, 46, 54, 58, 66, 67, 77, 178, 195, 238
resource, 18, 27–29, 31–36, 42
scheduling, 56–58, 63, 65, 75, 81
self-organizing maps, 158, 163, 165
Semantic Web, 25, 26, 31–34, 42–44, 46, 47, 80

singularity, 62, 66, 68, 72, 75
Snakemake, 59, 64–68, 72, 74, 75
sparse, 173, 178, 197, 200–204, 207, 209, 212, 215, 223, 224, 226–229, 232–235, 237, 244–246
statistical tests, 92, 93, 95, 125, 126, 132
subsumption, 31, 32, 36, 39, 40
Taverna, 59, 64–66
terminologies, 3, 5, 7, 9, 17, 198, 199, 207
traceability, 18, 19, 61, 63, 81

## U, V, W

URI (Uniform Resource Identifier), 31–34, 43
variable selection, 89, 92, 99, 108, 109, 111, 167, 203, 211, 216, 233
variance, 91, 94, 108, 110, 127, 128, 130, 132, 153, 162, 164, 169, 170, 179, 181, 198, 199, 201–204, 207, 210–212, 224, 227
variant, 16, 119, 120, 124, 126–133, 137, 164, 197, 203, 207, 208, 212
weight, 125, 129, 130, 132, 136, 138, 157, 167, 169, 171, 173, 197–200, 209, 216, 221
workflow, 19, 44, 53, 54, 59–69, 72–81
   repositories, 61, 79
wrappers, 66